Remembering Scottsboro

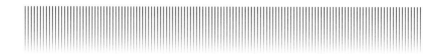

Remembering Scottsboro

The Legacy of an Infamous Trial

James A. Miller

PRINCETON UNIVERSITY PRESS

PRINCETON AND OXFORD

Contents

Illustrations

Acknowledgments

This book goes back a long way and has its origin in a larger project on African American cultural politics of the 1930s, initiated at the W.E.B. DuBois Institute for Afro-American Research in 1993; I thank the Institute, its Director Henry Louis Gates, Jr., its staff, and my colleagues at the Institute for their support. I continued my research as a Fellow of the Schomburg Center's Scholar-in-Residence Program, funded by the National Endowment for the Humanities/Aaron Diamond Foundation. I thank both the staff and my colleagues at the Schomburg for their support; a particular note of thanks goes to Diana Lachatanere. This work also owes significant debts to the joint research launched by my friends and colleagues Susan Pennybacker and Eve Rosenhaft, based in part on research Susan had conducted in the Russian State Archives of Social and Political History in Moscow—research that led to our jointly authored article "Mother Ada Wright and the International Campaign to Free the Scottsboro Boys, 1931–1934." That article, in turn, was an outgrowth of "Images of Scottsboro: Racial Politics and Internationalism in the 1930s," a paper coauthored by Susan and me and that we presented at "Racializing Class, Classifying Race: A Conference on Labour and Difference in Africa, USA and Britain," St. Antony's College, Oxford, in 1997. Susan, Eve, and I also presented some of our joint research on the international dimensions of the Scottsboro case at the University of Manchester, England, in 1998, and at the Third International Conference of the Collegium for African American Research (CAAR) in Münster, Germany, and at the Anglo-American Conference, University of London, in 1999. I thank Susan for her generosity in sharing the resources of her private files.

I am grateful for the assistance from the staffs of the following libraries and collections: Alabama Department of Archives and History, Montgomery, Alabama; Beinecke Library, Yale University; Manuscript, Archives, and Rare Book Library, Emory University; Library of Congress; National Archives, College Park, MD; Southern California Library for Social Studies and Research, Los Angeles; Moorland-Spingarn Research Center, Howard University; Schomburg Center for Research in Black Culture, New York Public Library; the New York Public Library for the Performing Arts; Tamiment Library, New York University; and the Special Collections Research Library, Syracuse University Library. I also appreciate the assistance of Paul Gardullo of the National

Museum of African American History and Culture, Smithsonian Institution, and Wendy Wick Reaves and Lizanne Garrett of the National Portrait Gallery, Smithsonian Institution. And I owe a special note of thanks to Elisa Marquez of AP Images.

I presented portions of this manuscript-in-progress at the 1998 annual meeting of the Japanese Association for American Studies, Chiba University, Japan; as the 1998 Tag Lecture at East Carolina University, Greenville, North Carolina; at the 2004 "African American Identity Travels Conference," University of Maryland, College Park; and at Minsk State Linguistics University, Minsk, Belarus; Rutgers University-Camden; Auburn Avenue Research Library on African American Culture and History, Atlanta, Georgia; Wayne State College, Wayne, Nebraska; and Franklin and Marshall College, Lancaster, Pennsylvania. I am grateful to the audiences who heard and responded to this material.

And I am thankful for the support of my colleagues in the English and American Studies departments, and the Women's Studies Program, at the George Washington University: Chris Sten, Faye Moskowitz, Dan Moshenburg, Gayle Wald, Patrick Cook, Jennifer James, Patty Chu, Tony Lopez, John Vlach, Phyllis Palmer, Melani McAlister, and Jim Horton. I also deeply appreciate the support of the office of the Dean of Faculty of Columbian College of Arts and Sciences.

Numerous friends and colleagues across the country offered information and insight as I worked on this project and I thank all of them, especially the late Lillian Robinson, Howard Gillette, Paul Lauter, Gary Okihiro, Maurice Jackson, James C. Hall, Van Gosse, Eliza Reilly, Alan Wald, Barbara Foley, William Maxwell, James Smethurst, Rick Hornung, Nicholas Moschovakis, Joanne Van Tuyl, Jacquelyn Dowd Hall, Dianne McWhorter, Dan Carter, and Gene Solon. My old friend Bob Gore has been a constant source of support and insight. Maureen Kentoff deserves a special word of thanks for helping to maintain an atmosphere of sanity in my workspace.

For their wonderful meals and intellectual nourishment I thank my good friends in Books98: Kent Benjamin, Chuck Lawrence and Mari Matsuda, Shirley Parry, Ginger and Jerry Patterson, Domy Raymond, Mary Helen Washington, Yvette Washington Irving, and Sherry Weaver. In many respects they have functioned as my ideal readers.

Edjohnetta Fowler Miller, Ayisha K. Miller, and John W. Miller deserve special thanks for their encouragement and support throughout this project.

I am indebted to Thomas LeBien for launching this work. Brigitta van Rheinberg, now Editor-in-Chief of Princeton University Press, patiently shepherded it through its early stages; Clara Platter brought it to a successful conclusion. I thank Brigitta, Clara, and their excellent staff, particularly Ellen Foos, Donna Goetz, and Jodi Beder, for their attention and guidance at every stage of the editorial process.

Remembering Scottsboro

INTRODUCTION

||||||||||||||||||||||||||||||||||||||||||||||||||||||||||||||||||||||||||||||||||||||||||||||||||

If you were alive at any time between March 25, 1931 and June 8, 1950, you lived, willy-nilly, in a Scottsboro world. The air you breathed was affected; the speech you heard; the newspaper you read; your political and international outlook, no matter where you sojourned, on the so-called civilized earth. The present tensions between East and West owe much of their early growth to the gigantic morality drama which did one-night stands around the globe — in which American democracy was depicted as the hypocritical Ogre of Evil, and Somebody Else as Helper of Justice — in foundation of the pleas for financial help "to save the Scottsboro boys."

— John Lovell, "Review of Allan K. Chalmers's *They Shall Be Free*"[1]

America free Tom Mooney
America save the Spanish Loyalists
America Sacco & Vanzetti must not die.
America I am the Scottsboro boys. . . .

— Allen Ginsberg, "America"[2]

THE GHOSTS OF THE MODERN CIVIL RIGHTS MOVEMENT continue to haunt American life. While the 2001 trial and conviction of Thomas Blanton, Jr., and the 2002 conviction of Bobby Frank Cherry, for the 1963 bombing of Birmingham's Sixteenth Street Baptist Church, killing four young black women, seems to have brought this episode to a successful conclusion, other cases are still pending. Recent years have witnessed a new investigation into the 1955 kidnapping, torture, and murder of Emmett Till and the revival of the investigation into the infamous murders of three civil rights workers, Michael Schwerner, Andrew Goodman, and James Chaney, in Neshoba County, Mississippi in 1964. In Georgia, activists, some public officials and relatives of the victims have pressed for authorities to bring charges in the 1946 incident when a white mob pulled four black sharecroppers from a car near the banks of the Apalachee River, dragged them down a wagon trail, and shot them to death.[3] Unfortunately, neither the tangled racial history nor the political culture of the United States is conducive to anything approaching a national "Truth and Reconciliation Commission" about such outrages—in spite of the earnest efforts of the Clinton administration to orchestrate a series of "National Conversations on Race" during the late 1990s—so the litany of racial grievances will undoubtedly continue into the foreseeable future. In the midst of this ongoing process of historical reconstruction and legal redress, the infamous Scottsboro case of the 1930s continues to function—albeit intermittently—as a broad signifier for the history of American racial atrocities.

In April 2004, for example, Courtland Milloy, an African American columnist for the *Washington Post*, reflected on the recent arrest of three Howard County, Maryland black teenagers who had been arrested and jailed after being accused of raping a fifteen-year-old white girl in a high-school restroom. The young men were subsequently freed six days later and the Howard County State's Attorney announced plans to seek dismissal of all charges against them, but Milloy saw in this event the disturbing signs of a persistent racial mythology that recalled the notorious Scottsboro case of 1931: "one of the greatest travesties of justice in 20th century America."[4] Linking this episode to the 2003 conviction and sentencing, in Rome, Georgia, of a black high school senior, eighteen-year-old Marcus Dixon, to ten years in prison for having sex with a fifteen-year-old white classmate; and to the then pending trial of the celebrated basketball star Kobe Bryant, who was facing the possibility of life in prison if he were convicted of raping a white woman staff member at an exclusive hotel in Colorado, Milloy invoked the Scottsboro case as "a cautionary tale about the pitfalls that racial and sexual mythology hold for black men. Unfortunately, he went on to moralize, "some young black men have been unaware of that history and, as the saying goes, were doomed to repeat it."[5]

Milloy's rhetoric is symptomatic of a wider pattern of discourse which continues to circulate—or, more accurately perhaps, flare up—around public events rooted in the still explosive alchemy of race, sex, and violence in American life. The Scottsboro case has been routinely invoked by contemporary writers in wide, and sometimes bewildering, contexts. In a rather garrulous letter to the editor of the *Boston Globe* regarding the racial ancestry of golfer Tiger Woods, David L. Evans invoked Scottsboro to account for the existence of a particular form of solidarity across class lines in the black community: "My reference is a vestigial habit from the Jim Crow era when a person with only one drop of black blood was black, regardless of his other heritage(s). That Draconian policy sealed a fraternal bond among African-Americans as disparate as the poor, dark-skinned Scottsboro Boys of Alabama and the erudite, light-skinned Walter White of the NAACP."[6] When four Morehouse

College students were accused of raping a Spellman student, the lawyer of one of the defendants hyperbolically compared their case to that of the Scottsboro Boys, with the hint that it deserved a comparable international outcry: "this case reminds him of the Scottsboro Boys, nine young Black men who in 1931 were falsely accused of raping two White girls. Eight received death sentences and one received life in prison. Their case inspired a successful international campaign."[7] In the aftermath of the infamous O. J. Simpson trial, scholars Abigail Thernstrom and Henry D. Fetter, alluding to a reference poet Nikki Giovanni made to the O.J. Simpson trial as "Scottsboro redux," sought to set the historical record straight by attempting to systematically dismantle such "ludicrous comparisons."[8] During the late 1990s, a poster advertising a Washington, DC meeting of the International Socialist Organization to demand freedom for the celebrated death-row prisoner Mumia Abu-Jamal foregrounded a picture of the "Scottsboro Nine" with their defense attorney, Samuel Leibowitz, inviting its readers to "join the ISO's weekly meeting to discuss the lessons of the Scottsboro case and how the tactics of 1931 can be used in the fight against the racist death penalty today!"[9]

The Scottsboro case left a distinct imprint upon the generation that came of age during the 1930s, as even a cursory survey of biographies and memoirs of blacks and whites, across a wide spectrum of social and political backgrounds and from different regions of the country, makes clear.[10] Even more striking, however, is the way that Scottsboro has continued to function as a multivalenced reference in contemporary American life.

Why does this infamous case continue to resonate in American culture more than seventy years after it occurred and more than a quarter of a century after the last chapter of the case apparently came to a close? Why does the invocation of Scottsboro invite knowing looks and the assumption of shared understandings about the nature of racism and the workings of the criminal justice system in the United States—even when anecdotal evidence suggests that people attach significantly different meanings to the case? What are the cultural and historical forces that have kept the memories of Scottsboro alive—and, more to the point perhaps, what is the content of those memories? And what are the implications of the ways the Scottsboro case has been constructed over time for how Americans perceive and talk about race?

Remembering Scottsboro explores the ways in which the case—arguably the most celebrated racial spectacle of twentieth-century American history, at least up to the 1955 murder of Emmett Till—entered American daily life, providing a vocabulary and frame of reference that continues to serve a purpose even when its referents are no longer clearly visible. From the very outset of the case in 1931, against the backdrop of heated and contentious struggles among the International Labor Defense (ILD), the Communist Party, and the National Association for the Advancement of Colored People, a carefully crafted "Scottsboro Narrative" emerged that had significant impact on the ways the case was perceived and understood in the national and international arenas. Other, competing narratives emerged as social and historical conditions changed, from the 1930s to the 1970s—and it was not until 1976, when Clarence Norris, "the last of the Scottsboro Boys," was officially pardoned by George Wallace, the Governor of the state of Alabama, that the case can be said to have officially come to an end.

In the 1930s, through the efforts of the ILD, the Communist Party, and their supporters, accounts of the case were marked by a very strong sense of advocacy for the Scottsboro Boys; in the years following World War II, coincident with the onset of the Cold War and its far-reaching effects upon American culture, accounts of the Scottsboro case were often inflected with sharply anti-communist rhetoric that persistently cast doubt upon the underlying motives of the Communist Party's involvement in the case. The emergence of the modern Civil Rights Movement in the late 1950s signaled the appearance of a Scottsboro Narrative that often placed the moral and political dilemmas of white subjects at its center.[11] Taking the 1930s "Scottsboro Narrative" as its point of departure, Remembering Scottsboro tracks its construction, its disaggregation, its reconstitution, and its sublimation in journalism, poetry, fiction, drama, and film as it traveled through more than a half century of American life. Chapter 1, "Framing the Scottsboro Boys," revisits the fierce combat between the International Labor Defense (ILD) and the National Association for the Advancement of Colored People (NAACP) over the right to represent the Scottsboro Boys. The ILD handily won both the struggle for legal representation and the battle for the control of their cultural and political representation; that is to say, the ILD and Communist Party propaganda shaped the dominant visual, literary, and popular images of the Scottsboro Boys during the 1930s. These images were constructed very quickly, within weeks of the conviction of the defendants, and their death sentences, in April 1931. Drawing upon materials from the ILD files, the NAACP files, and internal memoranda of the Communist Party–USA, this examination offers a glimpse at the behind-the-scenes discussions, bickering, and debates that helped to shape the publicity about the case. The intense preoccupation among the Scottsboro defenders about race, class, gender—and, more specifically, deeply entrenched attitudes and stereotypes about black masculinity—all fueled the rhetoric generated by the case. "Framing the Scottsboro Boys" meditates upon how this rhetoric created new opportunities to mobilize favorable public opinion at the same time that it foreclosed other possibilities of imagining and constructing the "boys." It also delineates the emergence of two significant and sometimes overlapping counternarratives to that proposed by the Communist Party, one largely shaped by Walter White and the NAACP, the other by southern apologists, some of whom seemed genuinely taken aback that critics of the state of Alabama did not give it sufficient credit for the humanity and restraint of its citizens who, after all, had not lynched the

Scottsboro Boys on the spot and had allowed justice—such as it was—to run its course.

Chapter 2, "Scottsboro, Too: The Writer as Witness," tracks the trajectory of the "Scottsboro Pilgrimage" as a distinct feature of the emerging rhetoric of the Scottsboro campaign in the early 1930s. Beginning with the poet and writer Langston Hughes, who visited the Scottsboro Boys during his first reading tour of the South in 1932, a number of poets and writers traveled to Alabama either to visit the defendants or to bear witness to the series of trials they endured. Often appropriating the Christian motif of bearing witness to the suffering of Christ on the cross, many of these writers—among them Muriel Rukeyser, Mary Heaton Vorse, John Hammond, and Louise Patterson—hoped to mobilize a movement of social justice by giving literary expression to the bodily suffering of the Scottsboro Boys—producing an extensive body of poetry, essays, and reportage based upon their journeys to Alabama and the South.

Contemporary plays and, in several cases, films were also inspired by the Scottsboro case. John Wexley's *They Shall Not Die*, Paul Peters's *Stevedore*, Langston Hughes's *Scottsboro Limited* all took the Scottsboro case (or roughly comparable incidents) as their point of departure; films such as Fritz Lang's *Fury* (1936) and Mervyn Leroy's *They Won't Forget* (1937) explored the issues of aggrieved innocence and trial-by-mob rule which many saw as the heart of the Scottsboro case, while William Wellman's *Wild Boys of the Road* in effect rewrote the Scottsboro case, turning it into a Depression-era "road" saga. Chapter 3, "Staging Scottsboro," examines the ways in which key playwrights and filmmakers joined the chorus of Scottsboro defenders during the 1930s.

Novelists weighed in on the case as well. Novels like Grace Lumpkin's *A Sign for Cain* placed the false accusation of rape at the heart of her somewhat melodramatic anatomy of life in a small southern town. Arna Bontemps's *Black Thunder* and Guy Endore's *Babouk* were both novels ostensibly concerned with significant but historically remote slave rebellions, but contemporary readers would have been quick to recognize the ways in which these novels both masked and expressed some of the urgent political concerns and debates of the 1930s. Chapter 4, "Fictional Scottsboros," analyzes the ways in which writers mined different literary genres in their efforts to bring Scottsboro to the attention of the reading public.

Focusing largely on the life and work of the most prominent African American writer of his times, Chapter 5, "Richard Wright's Scottsboro of the Imagination," considers the ways in which Wright's position as a cultural ambassador, a bridge between the black community and white readers, brought the struggles and terrors of African American life into a larger public arena. At the heart of Wright's sharp critique of Jim Crow racism and its corrosive impact upon the African American psyche stands the inextricable link between sexual terror and lynching—for which the Scottsboro case is often the signifier.

This chapter mines Wright's novels and his early poetry and short fiction to probe how his deepest preoccupations shaped his writing. This chapter concludes with a brief consideration of William Demby's important but neglected 1950 novel, *Bettlecreek*—in which the Scottsboro Narrative is recast with a white victim at its center, signaling an important shift in the version of the Scottsboro case that had circulated during the past two decades.

Chapter 6, "The Scottsboro Defendant as Proto-Revolutionary: Haywood Patterson," focuses upon the most uncompromising and truculent of the Scottsboro defendants, who attracted both the hostility of the Alabama prison authorities and the attention of the Scottsboro defenders, many of whom invested their hopes in the revolutionary potential of the black masses in his presence. Somewhat like Richard Wright's fictional Bigger Thomas, Patterson emerged as a wily and enigmatic personality who eluded the categories and definitions ascribed to him.

Chapter 7, "Cold War Scottsboros," follows the trajectory of the Scottsboro Narrative through the peak years of the McCarthy era, examining the ways in which the basic elements of the case were recast and reconstituted during the 1950s, and setting the stage for what I see as the final chapter of the Scottsboro Narrative, Harper Lee's *To Kill a Mockingbird*, to which I turn in Chapter 8.

In many respects this is an archival project. My purpose has not been primarily to appeal to a sense of retrospective indignation about a particularly sordid episode in twentieth-century American racial history (although, to be sure, any sober account of the case invariably provokes such a reaction), but to explore the ways in which the shifting lexicon surrounding the Scottsboro case sheds light upon shifting and enduring American attitudes towards race and justice.

||||||||||||||||||||||||||||||||||||||||||||||||||||||||||||||||||||||||||||||||||||||||||||||||||||||||||||||||||||||||||||||||||||||||||||||

CHAPTER ONE

FRAMING THE SCOTTSBORO BOYS

HAD IDA B. WELLS-BARNETT—the fearless and uncompromising "crusader for justice," who had almost single-handedly launched the campaign against lynching in the late-nineteenth-century United States—been in her prime, she undoubtedly would have sprung into action as soon as the Scottsboro case broke. Ironically, on March 25, 1931, the same day that Wells-Barnett died of uremic poisoning in Chicago, the nine black young men who subsequently became known as the Scottsboro Boys were apprehended in Paint Rock, Alabama, charged first with vagrancy, and subsequently with the capital crime of rape. Wells-Barnett would have immediately recognized the Scottsboro case as a particularly egregious eruption of the problem that had galvanized her attention as a journalist and launched her career as a political activist almost forty years earlier; and she would have been quick to grasp the peculiarities of the case as it unfolded, particularly the role that the court system of the state of Alabama played in meting out penalties that, more often than not, had been exacted on the bodies of black men by violent lynch mobs. Unfortunately, some of Wells-Barnett's contemporaries, particularly those who claimed the responsibility of representing the rights and interests of the black community—most notably the National Association for the Advancement of Colored People (some members of whom had played an underhanded role in maneuvering Wells off of the "Committee of Forty," the group charged with defining the structure and priorities of the fledgling organization)[1]—did not immediately grasp the significance of Scottsboro. This crucial lapse of attention during the early stages of the ordeal of the Scottsboro Boys helped set the stage for one of the most highly visible—and carefully orchestrated—racial spectacles in the decades preceding the emergence of the modern Civil Rights Movement.

What was it about the Scottsboro Boys in particular that commanded such instant and widespread attention and, in some quarters, sympathy? The now classic UPI/Bettman photograph, widely published in newspapers across the United States on March 26, 1931, effectively, if unintentionally, dramatizes their plight. Literally and symbolically framed by eight members of the Alabama National Guard, four of them armed with rifles and bayonets, the nine prisoners convey a wide range of attitudes through their postures: Clarence

Figure 1.1 "Young Black Men Accused in the Scottsboro Rape Case." Source: Bettman/CORBIS (March 1931).

Norris, hat in hand, stands erect—almost military fashion—and stares directly at the camera; Olen Montgomery stands in a similar fashion, but his lips are parted and his eyes are downcast and appear swollen—a function, no doubt, of his blindness in one eye and poor vision in the other. Behind Montgomery, Andy Wright looks directly at the camera with narrowed eyes. Willie Roberson leans to one side, possibly to lessen the pain of the advanced case of syphilis that afflicts him. Ozie Powell and Charlie Weems stand in the back of the photograph, their images framed by a triangle of bayonets. Eugene Williams and Roy Wright—the two youngest Scottsboro Boys—look querulous and uncertain. Slightly to their right stands Haywood Patterson, hands thrust in his pockets, feet spread apart; he already conveys the attitude of truculence and defiance that would make him the most obvious target for the hostility and wrath of those people who wanted to take the lives of the Scottsboro Boys—and the symbol of defiance for those supporters of the boys who sought to find revolutionary meaning in their plight. All nine of them, however, appear weary and wary, traumatized—like the survivors of an ambush.

In the context of the racial terror endemic in the South during the 1930s the Scottsboro case was far from the worst atrocity visited upon black people. Yet the magnitude of the case, the sheer number of defendants, their relative youth and bewilderment, the atmosphere of carnival and menace that surrounded them from the very beginning, the rapidity of their trials, and the swiftness and viciousness with which death sentences were imposed upon eight of them—all of these factors conspired to elevate Scottsboro to the status of a modern morality play, a racial spectacle that conformed almost exactly to Grace Hale's clinical description of the spectacle of lynching:

> Over time lynching spectacles evolved a well-known structure, a sequence and pace of events that southerners came to understand as standard. The well-choreographed spectacle opened with a chase or a jail attack, followed rapidly by the public identification of the captured African-American by the alleged white victim or victim's relatives, announcements of the upcoming event to draw the crowd, and selection and preparation of the site. The main event then began with a period of mutilation—often including emasculation—and torture to extract confessions and entertain the crowd, and built to a climax of slow burning, hanging, and/or shooting to complete the killing. The finale consisted of frenzied souvenir gathering and display of the body and the collected parts.[2]

The Scottsboro case began with the "chase" of the slow-moving freight train from Stevenson, Alabama—where news of the fight between blacks and whites on the train was first reported—to Paint Rock. The Deputy Sheriff of Paint Rock informally deputized every white man in town who owned a gun, thus contributing to the highly charged atmosphere that greeted the young men. Loaded on an open flatbed truck, the nine black men were paraded from Paint Rock to the jail in Scottsboro, twenty miles away; there, six of them were identified by Victoria Price as the men who had raped her. Once the accusations of Victoria Price and Ruby Bates began to circulate, a crowd of several hundred white men began gathering in Scottsboro—a dress rehearsal for the reported crowd of ten thousand that converged on Scottsboro on the opening day of the trials, Monday, April 6, 1931.

What distinguished the spectacle of Scottsboro from the "classic" lynching scenario was the fact that the "main event"—the public meting out of "justice" and the "period of mutilation"—occurred symbolically within the courtroom (although acts of violence were also inflicted upon some of the defendants offstage, so to speak—in the bayonet attack by a national guardsman on Clarence Norris after he vehemently and publicly called Victoria Price and Ruby Bates liars the day after the boys were arrested; and in the vicious beating Norris suffered at the hands of prison guards the night before the first day of the trials—when he turned state's evidence for the prosecution). In this respect, the International Labor Defense (ILD) and the American Communist Party aptly designated the Scottsboro trials as a "legal lynching," and they fully understood that their defense of the Scottsboro Boys would depend, in part, upon their ability to disrupt a narrative that had, for many decades, made the spectacle of lynching seem natural.

Hale notes, too, another distinguishing feature of the modern lynching spectacle:

> The majority of Americans—white and black, northern and southern— learned about these events from newspapers and to a lesser extent books, pamphlets, and radio announcements. In many cases these accounts were written by reporters who personally witnessed the spectacle, but the experience for their readers or listeners was mediated, a representation at least once removed from actual involvement. And even those spectators who attended the lynching . . . were affected as well by the narratives constructed by reporters to describe and explain these events.[3]

In other words, as a result of decades of the social ritual of lynching a vocabulary and a narrative had emerged that standardized the perspective of "witnesses" across racial and political divides. This certainly was the case in the earliest reports of the Scottsboro case in the southern press. The March 26, 1931 headline of the *Chattanooga News*, for example, implicitly applauded the "restraint" of the mob gathered at Scottsboro—"QUIET REIGNS AS 9 NEGROES HELD IN ASSAULT CASE"—at the same time that it relegated the boys themselves to a familiar position in the hierarchy of racial stereotypes: "while the crowds gathered outside, the negroes [sic], ranging from 15 to 25 years in age, slept in the jail seemingly little concerned."[4]

From this perspective the UPI/Bettman photograph simply confirm℮ reinforced what many people already believed: the image of nine dish and bewildered young black men surrounded—literally framed—by stern, authoritative white men could certainly be read in terms of the absolute necessity of restraining and containing them. But those people whose social and political passions were aroused by the predicament of the Scottsboro Boys knew, or believed, that the "classic" narrative of spectacle lynchings would have to be challenged, subverted, overthrown—that an effective counternarrative would have to be constructed—if there was to be any hope of achieving justice for them.

Given the climate of racial and sexual hysteria incited by the southern press, it is striking that the stories that provided the basis for a counternarrative to the "lynching spectacle" were provided by two white women, neither of whom was from the South: Helen Marcy and Hollace Ransdall.

"Helen Marcy" was the name adopted by Isabelle Allen, who, with her husband Sol Auerbach (better known in Communist Party circles as James S. Allen), had been assigned to organize in Chattanooga, Tennessee, where they began the publication of the *Southern Worker* in 1930.[5] Within a week of the arrest of the Scottsboro Boys, Marcy filed the first on-the-scenes coverage of the case in the Communist Party press. It was Marcy who first called into question the veracity of the rape accusations and alerted her readers to the possibility of a lynching occurring in Scottsboro, as she wrote in her April 4, 1931 article in the *Southern Worker*:

> Nine young Negro boys, charged with "forcefully ravaging" two white girls on a moving freight train, barely escaped wholesale lynching today, but a certain lynching is being prepared for the date of the trial, April 6th, which is horse-swapping and general fair day here, when about 1000 farmers will be in town, incited to lynch fever by local merchants and the newspapers.[6]

Marcy hammered away at the lynching issue, drawing a sharp distinction between the "bosses" and the "workers." Hence, she reported, the local attorneys appointed by the judge to defend the Scottsboro Boys would function, in effect, as surrogates of the bosses in a legal lynching, "if that farce is to be gone through"; "The Negroes were indicted immediately with no lawyers present and without a chance to explain or defend themselves. Boss justice works quickly against the workers"; a mob spirit could be seen on the streets, "But, with many of the merchants absent, the crowd had no leadership."

Most noticeably, Marcy's reportage sought to establish objective correlations between the plight of the farmers gathering in Scottsboro—"most of them dressed in rags, with no coats, in overalls patched a thousand times, with thin, emaciated faces, the result of a winter of Red Cross support"—and that of "equally starving and wretched Negro workers." And she ended her

Figure 1.2 Victoria Price and Ruby Bates. Source: Bettman/CORBIS (January 5, 1932).

article with a rousing call to action: "To prevent the certain lynching on April 5, the white workers and farmers of Scottsboro will have to get together with the Negroes and defend the nine youths."

Similarly, on the basis of her firsthand contacts with the Chattanooga families of the Scottsboro defendants—Ada Wright, Claude and Janie Patterson, and Mamie Williams—it was Marcy who quickly recognized the human, and strategic, value of enlisting the families as a key element in mobilizing popular sympathy and support for them. Marcy introduced an April 18, 1931 article in the *Southern Worker* with excerpts from a purported letter by Haywood Patterson:

My Dearest Sweet Mother and Father:—
This is to let you know of my present life and worried to think that your poor son is going to die for nothing.[7]

While Patterson's heartfelt expressions of love and affection for his parents, as well as his manner of youthful dependence and aggrieved innocence, framed the article, Marcy seized the opportunity to locate the Patterson and Wright families within the harsh economic conditions of Depression America, noting that they live in "clean, but very poor working class homes." Mr. Patterson, she notes, works in a steel mill for a "measly $7 for a family of eight," while Andy and Roy Wright lost their father seven years ago and worked wherever they could find jobs to support their mother, Ada, "who makes $6 a week for working day and night in a 'white folks' home." In the face of these grim economic circumstances, Roy and Andy Wright and Haywood Patterson hopped on a train heading for Memphis, hoping to find work that would ease the burden on their families. The result was the Scottsboro case and their impending execution: "They are in constant danger of being lynched. Their food and bedding is not fit for pigs. The bosses newspapers are all excited that they dared raise their voices against being electrocuted for a crime they did not commit"—and, Marcy concludes, thereby introducing an important subtext in the defense of the Scottsboro defendants—"on the word of two notorious prostitutes. . . . From inquiries in the neighborhood, I learned that the two white girls, Victoria Price and Ruby Bates, are well known as prostitutes in that section where they have plied their trade for a number of years."[8]

If Helen Marcy reported on Scottsboro from the perspective of a committed member of the American Communist Party, writer-activist Hollace Ransdall brought a somewhat more nuanced perspective to the case—even though she apparently shared Marcy's concerns. Ransdall, who could have easily provided a model for a character like Mary French in John Dos Passos's celebrated U.S.A. trilogy, was born in Colorado and educated at Colorado College. After graduation, Ransdall slowly made her way east, first to Chicago, where she took courses at the University of Chicago, then to New York City, where she studied journalism and economics at Columbia University, worked as a volunteer with the American Civil Liberties Union, and became actively involved with the Aid the Passaic Strikers Committee. She also began spending her summers working at the Southern Summer School for Women Workers in Industry. When the Scottsboro case broke, Forest Bailey, the director of the ACLU and a friend of Ransdall's, asked her to investigate the case on its behalf.[9]

Ransdall's expedition to Alabama just happened to coincide with the first visit of Walter White, the executive director of the NAACP, in his efforts to gain firsthand knowledge of the case. Years later, she could still recall having dinner with White and meeting him in his Birmingham hotel room for briefings on the case:

So I was to meet him at this hotel in Birmingham. Which I did. Walter was a Negro in blood, but he was white. And he got such a kick out of this because he said they would lynch me if they knew . . . he took me out to dinner that night and told me the whole background of the case and so forth and gave me all my start. And he got such a chuckle out of it. He kept laughing. What a joke on these people in the hotel! Invited me up to his room and we had a little conference in his room. . . . He had that kind of sardonic humor. . . . This was all very new to me and I was a little shy and I was a little afraid of him. [Laughter.] So he told me where the trial was and helped me get my ticket up to Scottsboro and I started off on my own. I was all right then, you know, when I was running it. When he was trying to help me out and taking me out to dinner and entertaining me in his hotel room and all this sort of business, I was out of my element. But when I was on my own after I got up to Scottsboro, then I knew what I was doing. Inquiring, going around asking the people who should know.[10]

For ten days in May Ransdall traveled around the Scottsboro area, interviewing public officials, judges, mayors, prosecutors, doctors, social workers, and townspeople. Most importantly, she gained the confidence of, and extensively interviewed, the accusers of the Scottsboro defendants, Ruby Bates and Victoria Price, and their mothers. Interestingly, neither the voices of the Scottsboro Boys nor their families figure prominently in Ransdall's report—although she briefly mentions a visit with them in their cells on death row on May 12, 1931. The result of her sojourn remains not only one of the best historical accounts of the circumstances of the case, but also a graphic portrayal of an encounter between the sensibilities of a young, northern, politically left-of-center white woman and the deeply entrenched racial and social attitudes of a small southern town during the early years of the Depression.

The first part of "Report on the Scottsboro, Ala. Case" carefully reviewed the history of the first Scottsboro trials, pointing out along the way—but without much editorial comment—clear discrepancies within the accounts provided by the court testimony of Victoria Price and Ruby Bates, such as the claim that they had spent the evening before the fateful train ride at the Chattanooga home of a Mrs. Callie Brochie—or the haziness of the two accusers about the number of black men supposedly involved in the rape on the train. Like Helen Marcy, Ransdall was on the alert for the presence of "a lynching spirit" in Scottsboro, but she was much more measured in her assessment of its possibility. Ransdall boldly and graphically reported the inconclusive nature of the sexual examinations of the two accusers and flatly repeated the testimony that "Willie Robeson [one of the Scottsboro defendants] was suffering from a bad case of venereal disease, which would have made it painful, if not impossible for him to have committed the act of which he was accused."[11]

As Ransdall surveyed the public record of the case, she must have realized that its foundations were so flimsy that the sources for the passions surrounding it would have to be found elsewhere.

Following Helen Marcy's reportage, Ransdall appropriated the rhetoric of the Scottsboro Boys as black youthful innocents threatened by white southern menace: "without even as much as the sight of one friendly face, these eight boys, little more then children, surrounded by white hatred and blind venomous prejudice, were sentenced to be killed in the electric chair. . . . It is no exaggeration certainly to call this a legal lynching."[12] And she described her May 12th visit to see them in similar terms: "frightened children caught in a terrible trap without understanding what it is all about."[13]

In her determination to learn what the Scottsboro case was really all about, Ransdall probed deeply beneath the surface of small-town southern charm for the answers. The real heart of her investigation revolved around four fundamental issues: why Victoria Price and Ruby Bates made the charge; Ruby Bates and her family; Victoria Price and her mother; and why the boys were hated.

Ransdall maintained that the first question could not be answered without reference to the social and economic conditions that governed Huntsville, Alabama, the hometown of the two accusers—a classic case of poor working conditions in old-fashioned mills, long hours, and low wages. "High standards of morality, of health, of sanitation, do not thrive under such conditions," Ransdall noted. "It is a rare mill family that is not touched in some form by prostitution, disease, prison, insane asylum, and drunkenness."[14] Family life is unstable under these conditions, she noted; husbands come and go.

Ransdall shrewdly allowed local social workers to judge these conditions, somewhat distancing herself from the judgments a northern outsider might make: "These mill workers are as bad as the Niggers," said one social worker with a mixture of contempt and understanding. "They haven't any sense of morality at all. Why, just lots of these women are nothing but prostitutes. They just about have to be, I reckon, for nobody could live on the wages they make, and that's the only other way of making money open to them."[15]

Having thus located Victoria Price and Ruby Bates at the bottom of the white social hierarchy of Huntsville, Ransdall clearly indicated her preference for Ruby Bates as the more appealing and compelling of the two accusers: "Ruby is only seventeen. She is a large, fresh, good-looking girl, shy, but a fluent enough talker when encouraged." Ruby lives in a "bare but clean unpainted shack," and Ransdall took sharp exception to the way her living conditions were characterized by the local social worker: "The social service worker who accompanied me on the visit sniffed when she came in and said to Mrs. Bates: "Niggers lived here before you, I smell them. You can't get rid of that nigger smell." Mrs. Bates looked apologetic and murmured that she had scrubbed the place down with soap and water. The house looked clean and orderly to me. I smelled nothing, but then I have only a northern nose."[16]

Noting the younger Bates children playing with their black neighbors, Ransdall meditated upon the cruel irony of the Scottsboro case:

If the nine youths on the freight car had been white, there would have been no Scottsboro case. The issue at stake was that of the inviolable separation of black men from white women. No chance to remind Negroes in terrible fashion that white women are farther away from them than the stars must be allowed to slip past. . . . As a symbol of the Untouchable White Woman, the Whites held high—Ruby. The Ruby who lived among the Negroes, whose family mixed with them; a daughter of what respectable Whites call "the lowest of the low," that is a White whom economic scarcity has forced across the great color barrier. All the things made the respectable people of Scottsboro insist that the Negro boys must die, had meant nothing in the life of Ruby Bates.[17]

Ironically, Ransdall's sympathetic portrayal of Ruby Bates and her apparent imperviousness to the magnitude of the political maelstrom she had, apparently unwittingly, helped to create would provide a vocabulary for Ruby's carefully crafted image in the hands of the American Communist Party after her "conversion" to their cause right before the Decatur Scottsboro trial in 1933.

If Ransdall conferred some degree of respectability upon Ruby Bates, she was very scornful of Victoria Price, beginning with her reports of the widespread uncertainty about her actual age: "Victoria lives in a little, unpainted shack . . . with her old, decrepit mother." She is " . . . a lively, talkative woman, cocky in manner and not bad to look at," but she also has a bad reputation in the neighborhood, "and has been in no end of scrapes with married men." Clearly, such a woman could not be trusted. At the very least, Ransdall argued, "the question of whether a monstrous penalty has not been exacted for an offense which the girls themselves feel to be slight, can certainly be raised."[18]

Finally, Ransdall turned her attention to the nagging question of why the Scottsboro Boys were the subjects of such intense vilification. Ransdall was truly struck by the natural beauty of Scottsboro, Alabama—"a charming southern village . . . situated in the midst of pleasant, rolling hills. . . . A feeling of peace and leisure is in the air. The people in the streets have easy kind faces and greet strangers as well as each other cordially."—and found it very difficult to reconcile her impressions with the attitudes she encountered right below the surface of these pleasant encounters: "It came as a shock . . . to see these pleasant faces stiffen, these laughing mouths grow narrow and sinister, those soft eyes become cold and hard because the question was mentioned of a fair trial for nine young Negroes. . . . They all wanted the Negroes killed as quickly as possible in a way that that would not bring disrepute upon the town. They therefore preferred a sentence of death by a judge to a sentence of death by a mob."[19] Thus did Hollace Ransdall lay bare a truth otherwise obscured by numerous trials, thousands of pages of court testimony, and two Supreme Court decisions.

One final question lingered for Ransdall: why society neglected the boys. "The question of why these nine young Negroes . . . have never been given a chance to be anything but the illiterate, jobless young itinerants they are, lies tied up with the whole problem of the denial of civil, social, and economic rights to the Negro in America."[20] The solution to these problems—so clear to Helen Marcy and her committed comrades who had dedicated themselves to organizing blacks and whites in the deep South—was presented in more modest terms by Ransdall in the conclusion of her report: "It can be answered only by a study of the discriminations practiced against the Negro in all phases of his life—educational, residential, economic segregation."

Despite its understated tone, Hollace Ransdall's "Report on the Scottsboro, Ala. Case" had an electrifying impact upon the office of the American Civil Liberties Union, and its effect rippled outward, helping to shape northern liberal and mainstream opinion, the leadership of the NAACP, and attitudes on the Left as well. Even in the midst of the fierce ideological contention that raged among the International Labor Defense, the American Communist Party, and the NAACP, Ransdall's report provided an analysis and a vocabulary, and the outlines of a script about the Scottsboro case, that could be appropriated by all sides of the ideological divide.

The attempt by the American Communist Party and the International Labor Defense to fashion such a narrative began almost immediately after reports about the arrests at Scottsboro began to circulate in the national press. Forged within the crucible of the grueling, bitter, vicious, no-holds-barred battle for control of the defense of the Scottsboro Boys that raged throughout 1931 and flared up for years afterwards (even after the ILD, the Communist Party, and the NAACP achieved a provisional peace with the formation of the Scottsboro Defense Committee in December 1935), these battles would irrevocably shape the public meaning of Scottsboro for many years to come.

The American Communist Party struck first with its headlines in the April 10, 1931 issue of the *Daily Worker*: "WORKERS, NEGROES UNITE! STOP THE 'LEGAL' LYNCHING OF NINE NEGRO BOYS IN ALABAMA!" Behind the self-assurance of this slogan, there was considerable behind-the-scenes wrangling between Tom Johnson, the CP district organizer in Chattanooga, and his embattled comrades in the region, and the CP Central Committee in New York, as they struggled to develop a lexicon that would define the official view of the case. Writing on behalf of the Central Committee, Clarence Hathaway initially wired to Johnson thirteen proposed slogans on issues ranging from the defense of the Scottsboro Boys, to the racial composition of their jury and the guards assigned to protect them, to demands for cash relief for "starving farmers and unemployed workers, Negroes

Figure 1.3 "Dreiser on Scottsboro." Cover of *Labor Defender*, June 1931.

Figure 1.4 "Don't Let Them Burn." Cover of *Labor Defender*, October 1932.

and whites equally," to the confiscation of land from the landlords, to the "right of self determination for the Negro people."[21]

In correspondence with the Central Committee throughout much of April 1931, Johnson—sometimes patiently, sometimes testily—tried to educate his northern comrades and superiors about southern life, sharply demurring over several key issues. Regarding the slogan "THESE BOYS ARE INOCENT [sic]; DEMAND THEIR IMMEDIATE RELEASE," for example, Johnson carefully explained why his fellow organizers had decided against it, urging that this slogan be deferred until they had completed their investigation:

> The stories of the boys were in some minor respects contradictory. It was the opinion of the comrades who interviewed them that further investigation— and we are sending a comrade to investigate—is necessary before we can come out openly with a statement as to their absolute inocence although this is of course our presumption.
>
> Farther the great publicity of the whole trial and of the preported confessions has created an idea in the minds of even many Negro workers that at least some of them may be involved in the alledged crime but certainly the trial was a framed-up lynch law from start to finish. With so many boys involved I do not think there is any contradiction in for the present rejection of the slogan of unconditional release and retaining the slogan of frame-up.[22]

Johnson also tried to explain why the slogan "DEMAND NEW TRIAL BEFORE JURY COMPOSED OF WORKERS AT LEAST HALF TO BE NEGROES TO EXPOSE THIS FRAMEUP" flew in the face of local and regional realities:

> A jury . . . will be picked in a capitalist court right on the edge of the black belt where racial atagionisms [sic] incited by the ruling class are the sharpest. You may be sure, and the Negro masses will also quite well realize, that any whites, workers or otherwise, selected for such a jury under local conditions will be white chauvinists and will vote for conviction. . . . [Remember] the mob of 10,000 who shouted for the blood of these Negroes in Scottsboro during the trail [sic]. These were not bosses, they were all misled half-starved white mountaineers and croppers. What chance would Negroes have with a jury made up of half whites selected from among this crowd? And this is the kind that will be selected.[23]

Johnson insisted that "a separate slogan be included dealing with the special incitment [sic] of the whites on this issue of 'attacking white women.' This is most important as it is the standard issue in the South and by all means should be given by the CC in its directives." And he registered his absolute opposition to the slogan "DEMAND THE CONFISCATION OF LAND FROM THE LANDLORDS FOR THE NEGRO AND WHITE

CROPPERS," strenuously arguing that "it is not realistic and does not conform to the present stage of the struggle."[24]

Finally, Johnson expressed his reservations about the slogan "STOP THE LYNCHING": "no lynching is immediately in prospect in this case. We propose 'PREPARE TO STOP ANY ATTEMPT TO LYNCH THESE NINE NGRO [sic] WORKING CLASS BOYS.'"[25]

Thus, step by step, a framework for understanding the Scottsboro case in its broader social, economic, and political context was crafted. Hathaway wielded his authority on behalf of the Central Committee with the subtlety of a sledgehammer. In an April 21, 1931 letter to Johnson he berated him, first about the attitudes of Allen Taub, one of the ILD's top attorneys who had been immediately dispatched to Chattanooga to assist with the case, and local organizer Douglas McKenzie, who had actively followed the trials from the beginning. Then he returned to the issue of the relevant slogans for the defense campaign; he seemed particularly exercised about Johnson's proposed alterations to the "STOP THE LYNCHING" slogan:

> You forget that we have characterized the whole procedure in Scottsboro as a legal lynching. When you state "that no lynching is immediately in prospect" in this case it appears to us that July 10th [Note: *the date on which eight of the Scottsboro boys were scheduled to be executed*] is quite immediate. To us it does not matter whether the Negroes are lynched by a crazed mob before July 10th or whether they are lynched in an Alabama penitentiary. The fact remains that lynching faces these boys on July 10th. Why then, you can ask, do we not formulate the slogan, "Stop the legal lynchings of these boys" rather than merely "lynching" as we do formulate it? We leave out the legal precisely because it is necessary to mobilize the masses against both their legal lynching as well as against the possibilities, and one might say even the probabilities of the lynch mob.[26]

In this way, by executive fiat, was the lexicon of the Scottsboro case (including the term "legal lynching") shaped by the Communist Party and the International Labor Defense, and the image of the Scottsboro Boys carefully constructed. In the pages of the official journal of the ILD, *Labor Defender*, William Patterson, a leading African American Communist, offered the following version of the events leading up to the arrest of the Scottsboro defendants:

> The Negro workers were guilty of seeking work during a period when it was more profitable for the bosses to starve the workers than to give them employment. They "hopped" a freight train bound for Scottsboro where they heard work might be secured. In the car [where] the Negro workers [were] grabbed were seven white workers and two white women dressed in overalls.

These workers were also unemployed workers, the victims also of the bosses insane form of government with its rewards to the workers of mass employment, starvation, sickness without insurance, prostitution, bloody wars and death.

But these white workers had learned from the lips of the bosses themselves the great hoax of the ruling class—the legend of "white supremacy." To them reared in the boss created atmosphere of race hatred bolstered by privileges of lynching, mob violence and indiscriminate terrorizing of Negroes, this myth of white supremacy seemed real. 10,000,000 white, starving unemployed workers also "proves" white supremacy.

The box car occupied by white workers was too good for "niggers." But the black workers fought for their lives. The girls were injured in the struggle—but not touched otherwise. The white workers telegraphed ahead their version of the affair and the bosses' police waited for the train to reach Scottsboro while the bosses' press whipped the town into lynch frenzy with its big headlined story of the "attack" by "nine burly black brutes" upon two white girls. A doctor's examination of the girls proved conclusively that they had not recently been touched sexually. Of course the vicious Negro baiting press of the landlord class concealed this fact.

Condemned to death, the Negro youths revolted in prison and fought with the guards until they were beaten unconscious.

Paul Peters, then the publicity director of the ILD, whose jointly written play *Stevedore* would in the mid-1930s graphically dramatize many of the issues underlying the Scottsboro case, sought to humanize two of the Scottsboro Boys—Roy and Andy Wright—and to underscore their credentials as workers, in an early interview with their mother, Ada, who would soon emerge in her own right as a forceful defender of her sons and the other Scottsboro defendants:[28]

> [Roy] used to carry meals for the men that work in the foundry. Sometimes he'd make a dollar a week. Then, when the foundry kind of shut down, Roy picked up old junk, and paper, and rags and iron, and tried to sell them to the junk man. He made perty good at that. Roy wasn't no slacker, no sir! He used to make three and maybe four dollars a week sometimes.
>
> Andy, he tried awful hard, but jobs was hard to get. Andy run a truck for a produce house downtown, and he was doing well enough. . . . Times was getting hard, and they was laying off the colored folks anyhow. So they fired Andy. . . .
>
> Then, one day, Andy said he was going down to Memphis. "Folks say I can get a job loading boats in Memphis, maw," he said. "I'm going down and send you back some money for to live on." And Roy, of course, he wanted to go along.[29]

Although articles such as Patterson's sometimes emphasized the Scottsboro Boys as symbols of resistance, the dominant tendency of Communist Party strategists, writers, and speakers was to portray them as hard-working, industrious, and caring young men—innocent workers who had been victimized by the vicious backwardness of southern life. This was the image that most often circulated in statements about them—or in statements issued in their name, such as the appeal "To the Young Workers of the World":

From the death cell here in Kilby prison, eight of us Scottsboro boys is writing this to you.

We been sentenced to die for something we ain't never done. Us poor boys been sentenced to burn up on the electric chair for the reason that we is workers—and the color of our skin is black. We is none of us older than 20. Two of us is 14 and one is 13 years old.

What we guilty of? Nothing but being out of a job. Nothing but looking for work. Our kinfolk was starving for food. We wanted to help them out. So we hopped a freight—just like any one of you workers might a done—to go down to Mobile to hunt work. We was taken off the train by a mob and framed up on rape charges.

At the trial they give us in Scottsboro we could hear the crowds outside yelling, "Lynch the niggers." We could see them toting these big shot-guns. Call 'at a fair trial?

Only ones helped us down here been the International Labor Defense and the League of Struggle for Negro Rights. We don't put no faith in the National Association for the Advancement of Colored People. They give some of us boys eats, to go against the other boys who talked for the I.L.D. But we wouldn't split nohow. We know our friends and our enemies.

Working class boys, we asks you to save us from being burnt up on the electric chair. We's only poor working class boys whose skin is black. We shouldn't die for that.

We hear about working class people holding meetings all over the world. We asks for more big meetings. It'll take a lot of big meetings to help the I.L.D. and L.S.N.R. to save us from the bossman down here.

Help us boys. We ain't done nothing wrong. We are only workers—like you are. Only our skin is black.[30]

If emphasizing the youth, innocence, and helplessness of the Scottsboro Boys constituted one cornerstone of the Communist Party's strategy, the other key component was its systematic and relentless vilification of the NAACP. In the context of the Communist International's "Third Period" of sharpened class antagonisms and heightened expectations of impending socialist revolution, the Communist Party designated the NAACP as a class-and-race traitor,

Figure 1.5 Portrait of Scottsboro Boys Clarence Norris and Haywood Patterson by African American artist Aaron Douglas, 1935. Source: Smithsonian Institution National Portrait Gallery.

targeting it with the same depths of hostility and contempt previously reserved for "reformists" and "social fascists" within the labor movement.[31] As James S. Allen, a key Party theoretician on the "Negro Question" who had been leading the Party's organizing efforts in Chattanooga when the Scottsboro case broke, proclaimed:

> The main political significance of the earlier stage of the Scottsboro struggle rested in the fight between revolutionary forces led by the Communist Party and reformist forces represented by the National Association for the Advancement of Colored People. The struggle went far beyond the ques-

tion of who shall carry through the legal defense of the Scottsboro boys. It was a struggle between two opposing class forces.[32]

Not only did the Communist Party give the Scottsboro case central coverage in its publications, it routinely castigated the NAACP for not speaking out about it.

While the Communist Party and the ILD were busily constructing the public face of the Scottsboro campaign, the NAACP was working just as actively behind the scenes. Approaching the case with its characteristic caution and circumspection, the NAACP was initially taken aback by the ferocity of the Communist Party's ideological offensive. Dr. P.A. Stephens, secretary of the Chattanooga Interdenominational Ministers Alliance, first sought assistance from Walter White, executive secretary of the NAACP, in a letter written on April 2, 1931.[33] White initially responded cautiously; like Tom Johnson and his comrades in Chattanooga, White wanted more assurance about the innocence of the boys, and asked for more facts about the case. He also seemed to be casting about in search of the legal precedents that might buttress an appeal. An April 9, 1931 letter to Stephens from the Special Legal Assistant at NAACP headquarters suggested that there might be similarities between the Scottsboro case and the Elaine, Arkansas riots trials—a legal battle spearheaded by the NAACP that culminated in a 1925 Supreme Court ruling that a trial conducted in a mob atmosphere is a violation of legal due process.[34] Still, the NAACP would not proceed without reviewing the transcript of the case. In mid-April, it had still not made a decision to enter the case, and public criticism about its lack of involvement was beginning to mount in the black community and the black press—as well as in the strident rhetoric of the Communist Party and the ILD. By April 20th a note of urgency had crept into White's correspondence with Stephens. In an air mail–special delivery letter White appealed to Stephens to persuade the Scottsboro Boys not to make any commitments about who should defend them:

> May I strongly urge upon your committee that it communicate with the eight boys who have been sentenced to death to urge them that they sign no agreement as to their defense or commit themselves in any way, pending opportunity by us to examine the transcript of testimony and to determine what action we shall be able to take in this matter.
>
> I must urge this as the International Labor Defense, a Communist organization, is seeking, as you doubtless know, to use this case for the purpose of making Communist propaganda. It has been our experience that it is impossible to cooperate with this group. The National Association for the Advancement of Colored People is interested solely in saving the lives of these boys if innocent and of securing a fairer and more impartial trial than was the case at Scottsboro even if they are guilty.[35]

Within two weeks of the conviction of the Scottsboro Boys, the battle lines between the International Labor Defense, the Communist Party, and the NAACP were sharply and apparently irrevocably drawn.

In the intense jockeying and arm-twisting that followed, it appeared that White's appeal had had an impact, when all nine of the defendants signed the following statement on April 23, 1931:

> We . . . wish it understood that we desire our defense to be left with the Interdenominational Ministers Alliance of Chattanooga, Tennessee, and that Messrs. Milo Moody and Stephen Roddy continue to act as our chief attorneys as they did at the trial in Scottsboro, Ala. Dr. H.G. Terrell, pastor of St. John M.E. Church is the only man we will advise with except Mr. Moody and Mr. Roddy. . . .
>
> None of us want to have anything to do with the Interdenominational [sic] Labor Defense or its representatives and have not had. We did not send for the Interdenominational Labor Defense Organization, or its attorneys, or others who may be connected with it. They came to see us and we told them to see Mr. Roddy. We do not want our parents or other relatives to employ them or have anything to do with them.[36]

In the meantime, unknown to Walter White, the field secretary of the NAACP, William Pickens—widely regarded as one of black America's most popular and effective orators—had written a letter on NAACP stationery to the *Daily Worker* praising the leading role the ILD had taken in the Scottsboro case. Pickens saluted the ILD for its battle "to prevent the judicial massacre of Negro youth in Alabama." "This is one occasion for every Negro who has intelligence enough to read, to send aid to you and to I.L.D." Enclosing a small check to aid the cause, Pickens added:

> In the present case the *Daily Worker* and the workers have moved, so far, more speedily and effectively than all other agencies put together. If you do not prevent Alabama from committing these horrible murders, you will at least educate working people, white and black, to the danger of division and the need for union. In either event it will be a victory for the workers.[37]

Pickens's letter was published on April 24, 1931—the same day that Stephens wrote White reporting the success of the mission to secure the signatures of the Scottsboro Boys—along with a reply to Pickens from Communist Party leader Robert Minor: "I cannot tell you how much joy it gave me to see you come into this fight. . . . Now I am hoping to hear your thunderous voice raised all over the land."[38]

Needless to say, the Communists effectively used Pickens's statements to continue to pummel the NAACP and to fan the rumors that Pickens had so infuriated the board of the NAACP that he was going to be discharged. The NAACP did not demand Pickens's resignation—although he was sharply rebuked by the board—but Walter White acknowledged in his correspondence that Pickens's remarks had done a great deal of damage:

> The Communists have raised Hell with their blustering, intemperate and almost insane threats. Pickens' letter is being broadcast and it has done us immense harm. Even some of our best friends interpret his letter as either proof that we are definitely linked up with the Communists or as an indication that we have fallen down on the job and a split in the Association has resulted. It will be a long time before we recover from that blow and I say this most advisedly.[39]

Prodded by the furious pace of events, the NAACP announced that it was entering the Scottsboro case and arranging for an appeal of the verdicts. On April 30, 1931 Walter White left for Chattanooga to lay the groundwork for their defense, triggering in the process a protracted and vituperative struggle with the Communists that consumed the rest of the year.[40]

Protecting the image and reputation of the NAACP was clearly one of White's concerns as he sought to navigate the shoals of the Scottsboro case, but there were other tactical and pragmatic issues as well. The NAACP *was* a reform organization that, in its twenty-odd years of existence, had relied upon time-tested practices of working through the established legal system to achieve equal rights and justice for the black community.

In the early stages of the Scottsboro campaign White's commitment to working within the system was firm, his faith in the good will of "decent" white southerners undiminished. As he remarked in a letter just before he left for Chattanooga:

> Our only chance of saving their lives is by getting a lawyer, preferably an Alabamian, whose standing is such as to help mobilize effectively the considerable sentiment which, we are informed, exists in Alabama among white people of the decent sort, that the boys are innocent.[41]

Immediately after he arrived in Alabama, however, White pronounced the Scottsboro case "by far the most tangled and ugly of them all." All of the boys had "exceedingly bad" reputations—with the exception of one, he reported— although, he added, "when one sees the terrible poverty and ignorance from which they come . . . one can readily understand why they should be what they are." Ruby Bates and Victoria Price suffered from even worse reputations, White continued: "They are notorious prostitutes and one of them . . . was arrested in a disorderly house in flagrante delicto with a colored man a few months ago." Given their squalid circumstances, he observed, "the parents of the boys, as may be expected, are frightfully ignorant and several of them have been taken in completely by the I.L.D." White concluded his extensive summary of the case by suggesting the following stark alternatives: "So, here is where we are. As I see it it would be suicidal for us to be tied up in any way with that outfit of lunatics—so far as this case is concerned—the Communists. Our choice is to withdraw completely from the case and make public fully our reasons for doing so—the Communists and their tactics, and the ignorance and stupidity of the boys' parents. Or, to get absolute control of the cases. . . ."[42]

Upon his return to New York City from Alabama in mid-May, White wrote to Olen Montgomery (he sent a similar letter to each of the defendants), spelling out what the NAACP was prepared to do to defend him:

> The Advancement Association is willing to put forth every legal and legitimate effort on your behalf that can possibly be put forth. No person can do any more, and if anyone tells you that he can do more, he is not telling the truth. It is most unfortunate for you that the International Labor Defense has, as Ozie Powell and Clarence Norris write me today, "only been agitating." We do not approve of the sending of threatening telegrams simply to get in the newspapers. . . .
>
> Besides being able to get the best lawyers in the United States, the Advancement Association has also for twenty-two years been defending, under the Constitution and the laws of the United States and of the various states, the constitutional rights of the Negro as a citizen and as a human being. I enclose a leaflet which tells of some of these cases which we have won. . . .
>
> Ozie Powell and Clarence Norris today write me that they want the N.A.A.C.P. to defend them, as they both feel that our organization will do everything in its power to help them. They also asked the International Labor Defense to keep hands off. May I, as a friend, urge you to take similar action promptly, and to write your parents that you do want the Advancement Association and no other to defend you.[43]

White also handed to the chastened William Pickens the assignment of securing the commitment of the Scottsboro Boys and their parents to the NAACP as their legal representative. In a May 12, 1931 letter to Pickens—his first correspondence with him since the appearance of Pickens's poorly timed letter to the *Daily Worker*—White spelled out, in effect, the blueprint for the NAACP's counterattack against the ILD and the Communists:

> It is a harsh thing to say, but there is ample evidence to justify these statements that the I.L.D. is obviously far more interested in making Communist propaganda out of the cases than in doing the wisest things to save the boys. The wildest and most intemperate statements had been made, which in large measure have alienated the reaction of a number of people in the South who were shocked at the trial given these boys at Scottsboro. For example, telegrams and letters demanding immediate release of the prisoners were sent the governor, the Sheriff at Birmingham, the Warden of Kilby

Prison and to other State officials. Many of these have contained veiled and open threats which have served, as you may well imagine, knowing the South as you do, to inflame sentiment against the boys to the point where it is freely said that either electrocution will be made certain or a mob will be formed to break into Kilby Prison to lynch the boys.

All manner of lies have been told, and especially about the N.A.A.C.P. The efforts of the I.L.D., as reported in their literature and especially in the *Daily Worker*, seem now to be chiefly devoted to attempted destruction or discredit of the N.A.A.C.P. . . . It is Communist strategy to try to destroy and discredit every organization except their own in order to clear the field for their own propaganda and activity.

Only the NAACP, White maintained, brought a dispassionate point of view to the case:

The one point at issue and the only one at Scottsboro is to save the lives of the boys who, I am convinced . . . are not guilty of the crime of rape, if they are guilty of anything. It happens that they are illiterate and their reputations are not particularly good. . . .

Unfortunately, this ignorance is even more marked in the parents than in the boys. They have been taken in by the flamboyant and impossible promises which the Communists have made to them and are, more unfortunately, of the type who believe whatever the white man says to them regardless of the character of the man or the nature of the statements, in preference to what the most reliable Negro might say. They are tools in the hands of irresponsible people.

Finally, White concluded, "I now come to the matter of your letter which appeared in the *Daily Worker*." Speaking "as a friend and not officially," White recounted the "enormous bewilderment" about Pickens and the NAACP that his letter had created. Although, White continued, he had made numerous efforts to explain that Pickens was not a Communist, "the publication of the letter has definitely created the feeling that the Association has gone Communist, or that it has fallen down on the job, as the Communists charge, or that there is a split in the Association, also as the Communists charge."

White offered Pickens the following "unsolicited suggestion": to prepare a public statement disavowing any disloyalty to the NAACP, clarifying the context in which his letter was written, and "repudiating the vicious usage which the Communists have made and are making of your letter." Almost as an afterthought, White added:

If you want to go a bit further it seems to me that it would be an excellent thing for you to express disapproval in unmistakable terms of the campaign of lies and vilification which the Communists are waging, not only against the N.A.A.C.P., but against the Interdenominational Ministers' Alliance and every organization and individual which does not submit to Communist dictation.

May I say to you as a friend that from my personal contacts and from the great volume of letters from branches and others which have poured into the National Office that some such action on your part seems to me to be absolutely necessary to clear up this situation which can be permanently harmful both to you and to the N.A.A.C.P. More important than either of these is the fact that the continuation of this campaign by the Communists may insure death to nine innocent boys.[44]

White's icy tone was unmistakable, as were his explicit instructions to Pickens to move quickly to repair the damage to himself, and to the NAACP, to hew to the line, in short. In spite of his morally righteous rhetoric about saving the Scottsboro Boys, White was clearly signaling to Pickens to launch an all-out counterattack against the Communists—and Pickens responded with the fervor of a new convert eager to secure his redemption, even though he still seemed to cling to the hope that somehow the NAACP and the ILD could collaborate on the defense of the Scottsboro Boys.

On the eve of Pickens's departure for Alabama, White issued explicit instructions:

The main job as we see it is to get definitely settled who, if any, of the boys and the parents of the boys want the N.A.A.C.P. to defend them. What we want is to know this definitely and to know it in such a way that there will not be a repetition of what has occurred already, namely, that they should sign whatever agreement is put before them—today one for the N.A.A.C.P., tomorrow one for the I.L.D., and the day after one for somebody else.

But White also wanted Pickens to emphasize the NAACP's altruism, in sharp contrast to the suspect motives and tactics of the Communists, and to warn the boys about the dire consequences, should they align themselves with the ILD:

Another vitally important thing is to show as clearly as it is possible for them to comprehend it just what is involved in this case—that the Association is not trying to get into the case simply to get in and that we are not doing it solely because of our interest in the boys themselves, but because of the fundamental principle involved affecting all colored people. I think, in justice to them, you can well point out that the presence of the Communists in the case, and especially because of their tactics of threats and their wild, intemperate and untrue statements, they are placing themselves in double jeopardy. I wish you would make it clear to them that they must make up their minds, once and for all, that if they are to be defended by the I.L.D., then we will have to wash our hands of the case and make public in full our reason for withdrawing, should that become necessary.

Finally, White warned Pickens, he would have to prepare himself to confront the appalling ignorance of the parents of the Scottsboro defendants:

I think I ought to warn you in advance that you have probably never before encountered such ignorance as you will find here. You will, unfortunately, also find, too, that some of those concerned belong to the type of colored people who will believe anything said by a white man, no matter what the character of the man or the statements, in preference to anything that may be said by any Negro, no matter how sincere or intelligent. In the final analysis, this last named circumstance is really the crux of the whole difficulty.[45]

The next day White fired off to Pickens an airmail letter containing a clipping from that day's *Daily Worker* as the most recent example of Communist perfidy, underscoring why it would be impossible to cooperate with them:

It seems to us that it is no longer a question of our cooperating with the I.L.D.—it is clear that the I.L.D. wants no cooperation.

It is actually clear now the motives back of their tactics. You pointed out one of them in your excellent piece to the colored press—namely, that it is the Communist strategy to destroy any organization except their own and the more effective the other organization is, the more determined and vicious is the attempt to discredit and destroy it.

This is especially so because of the Communists' determination to proselytize among Negroes on the theory that since the Negro is the most oppressed group in America he should be the most fertile field for revolutionary propaganda. Unfortunately they don't know the brother as we do.

Along the way, White made it clear that the NAACP thought it had found the legal precedent for the defense of the Scottsboro Boys in the Elaine, Arkansas riot case:

The cases themselves in their sheer brutality and disregard of human rights are most touching one. On the other hand, they are not any more difficult than any other cases the Association has fought and won. As a matter of fact, they are less difficult than, for example, the Arkansas Cases.

It is equally true that there is no broad legal principle to be established in the cases. The principle involved in them was established by us in the Arkansas Cases, Moore v. Dempsey, 261 U.S. 86, in ruling that a trial in a court dominated by mob influence is not due process of law. If the boys are freed it will be almost wholly on the basis of that decision won in 1925.

Finally, however, White returned to the real issue, as far as he was concerned: whether or not it was possible to work with the Communists.

To return to your question as to the possibility of both our and the I.L.D. having the cases. If it were possible we certainly would join hands and do anything to save the lives of the boys but . . . it is now quite evident that they will not permit such a course. Should some of the boys wish us to defend them, while others choose to remain with the I.L.D., we will, of course, furnish defense for those who wish us to aid them. Under the circumstances, however, I don't think it possible for the N.A.A.C.P. and the I.L.D. to defend the same boys.[46]

Pickens's first encounter with the Scottsboro Boys—a talk at Kilby Prison delivered to them in the presence of the NAACP-retained counsel from Alabama, Roderick Beddow; Warden Walls, the turnkeys and other Kilby Prison officials, and three local African Americans—skillfully integrated the themes White had articulated; his remarks should therefore be seen as an elaboration of NAACP attitudes about the Scottsboro defendants at the time:

> Young men . . . I know your bewilderment and your confusion, and your natural grasping after every straw: I know that some of you signed up for the National Association to take charge of your defense, when Mr. White was here, and I know that a few days ago all of you signed again with the Communists, who brought here all of your parents and sisters and aunts and bewildered relatives, to impress you and to make you sign. . . .
>
> Both Mr. Beddow and I, as well as other people in the country, believe that at least most of you, probably all of you, are innocent; otherwise he would certainly not be here, and maybe I would not be here. My organization has asked me to make perhaps our last statement to you, and later to your parents, if I can break through the Communist barricade and reach any of your parents. . . .
>
> I know the crowd that is offering to defend you; I know their methods and their motives. And I know Alabama and the south. And I know that if you get out of this "scrape" in an Alabama court, it will be due solely to the justice of that court, and entirely in spite of the methods and present activities of those who were here and made you sign up again the other day. Left to their methods and activities alone, you would certainly be electrocuted. I am talking plain,—it's time for it.
>
> I do not mean that the Communists want to harm you; I think that they really sympathize with you and would help you if they could. But I mean that their chief aim is to use you as a means for interesting colored people in the Communist Party—and in that they are likely to make you a great sacrifice in Alabama. We who know the Negro better, know that they will not much interest colored people in their politics by the methods they are using; but they think they will interest colored people, because they know such little about the American Negro.
>
> They have bewildered and amazed your poor parents and relatives; they have paid their fares to New York and other parts of the country, have put

them on platforms and in parades—all for purposes of their own, and not for the primary purpose of keeping you out of the electric chair. Some of your poor mamas have been shown sights and "kindnesses" they have never known before, and they think they are helping you, but every sensible black man and white man in the south, or anywhere else in the country, knows that a parade in New York and a harangue and a threat on one of its public squares, will have darned little effect in an Alabama court—unless it be a BAD effect.[47]

Pickens reported to White immediately following his visit:

Those poor, frightened, abused boys, Walter. . . . Those boys have been frightened and taken advantage of by the Communists [sic] crowd—and I shall always remember, as we left, the sight of several actually slumping weakly against the wall of the prison, in sheer terror, and saying sadly: 'Tell my folks to have those people to lay off and let your folk handle our cases.[48]

And he added the following hastily written note to the transcript of his talk:

Walter, if those Communist pig-heads succeed in sacrificing those lads, poor ignorant dumb-driven cattle that they are, they will certainly make me their eternal enemy.

 Their savage and immoral lying, revealed in the papers and documents you sent me, shows their real motives and caliber. I still hold that they are not enemies, at heart, of the boys. But they do not care a damned though what occurs to those boys or to the whole Negro race. Their own HOPE-LESS political advance is all that interests them. I mean their politics is HOPELESS for an indefinite time to come in the western world here.[49]

Pickens resumed his attack on the Communists in a subsequent report to White, contending that they were the major obstacle to a fair settlement of the case:

Roddy [the attorney hired by the Chattanooga ministers to represent the Scottsboro Boys] told me confidentially that Judge Hawkins told him before the hearings yesterday that the CHIEF ISSUE now is the REDS—that if they were out of it, it might be simpler; that he (the trial judge) did not really think the boys should be put to death, but that the Co/mmunists [sic] are more of an issue than are the FACTS of the case. God, what a mess, and a savage confession! But that is true. Can you not see from the clippings I sent you that Alabama is now getting ready to kill the boys just to "show 'em," and the question of the guilt or innocence of the boys is fast being lost sight of. . . . ?

He also continued to rail against the parents and relatives of the Scottsboro Boys:

> Now as to those parents and relatives—and maybe this will make you think like lying down and quitting: they are the densest and dumbest animals it has yet been my privilege to meet. But I have been exceedingly patient. I went first to the Patersons [sic], both husband and wife were there. I told them of the peril, and of the situation. . . . The mother was really broken down by what I showed her, and was ready to turn everything over to the Ass'n—but the less sensitive and less sensible father, said: "Give us your phone number and we will see you before you leave. Let us think it over. . . ."
>
> They were moved—the more intelligent mother especially.—But later when I got around to the mother of Williams in another part of the town. There was one of the reds, leading the Paterson couple like two dumb animals, jumping them off into his car and racing off to some place where the Williams woman had already been concealed to prevent any intelligence from reaching her. . . .
>
> Have you ever in your life seen dumb cattle so hopelessly entrapped? The woman, Mrs. Patterson, however gave me a despairing and appealing look as she passed by in company of this shirt-sleeved Communist ignoramus. I think her husband is taken in, and that the reds give them even money, as they do the boys in prison.[51]

In the correspondence between White and Pickens there is not the slightest sign of doubt that they could be mistaken in their judgments about the Scottsboro Boys and their relatives: they regarded them as simply too dumb to grasp the complexities of their predicament—and, therefore, as ignorant dupes of Communist manipulation. Secure in this knowledge, White and Pickens devoted considerably more time and energy trying to thwart the Communists than they did in cultivating relationships of trust and respect with the Scottsboro families.

The battle for the allegiance of the Scottsboro Boys continued through the summer, with an endless barrage of visits to Kilby Prison, arguments, promises, emotional appeals, threats, and dire warnings. By this time, too, the ILD had managed to enlist some of the Scottsboro mothers as potent weapons in speaking tours across the United States and abroad[52]—even as the NAACP vociferously complained that the Communists were inventing "Scottsboro mothers" as the need arose: "When the supply of 'mothers' was inadequate to cover such meetings—there are but five living mothers—substitutes were found. All over the country 'mothers' were produced. . . ."[53]

In early August 1931, Roderick Beddow, the Birmingham lawyer retained by the NAACP, reported that the loyalties of the Scottsboro Boys were sharply divided:

Mr. Beddow stated that he had visited the boys at Kilby Prison twice and that on both occasions the boys had been most insulting saying, "We do not want you. You are representing the capitalists and are just trying to get us electrocuted."

Only Ozie Powell, Clarence Norris and Charlie Weems still expressed the desire to have the N.A.A.C.P. and Mr. Beddow defend them. Olen Montgomery, Haywood Patterson, Eugene Williams and Andy Wright absolutely refused to talk with Mr. Beddow; also Willie Robinson (sic). . . . [54]

Walter White responded immediately by making another foray into Kilby Prison.

Upon his return to New York, he wrote to Claude Patterson, Haywood's father, indicating that the parents were the impediment to Haywood's willingness to be defended by the NAACP:

I have just returned to New York from a visit to the boys at Kilby Prison. While there your son said to me that he wanted to be defended by the N.A.A.C.P; that his only reason for staying with the I.L.D. was because his mother and father wanted him to. He asked me to write you and give him the facts in these cases so that you may be able to understand just what the situation is. [55]

After extolling the virtues of the law firm of Fort, Beddow, and Ray, praising Roderick Beddow as the "best criminal lawyer in Alabama and one of the finest in the South," White turned his attention to attacking his ideological demons:

The Communists, through the International Labor Defense, are seeking to make propaganda for their cause through these cases. Some of the leaders of that movement have even gone so far as to admit frankly that their ultimate desire is to destroy the American government and if in doing that the life of your boy must be sacrificed, this will be done. [56]

Repeating the by now familiar themes of the NAACP counterattack against the Communists, White proudly recounted the twenty-two-year history of its successes, reminding Mr. Patterson of the dubious track record of the ILD: "In contrast with the record of victories of the N.A.A.C.P. is the record of the International Labor Defense. This latter organization has not been able even to save white people. . . ."

In spite of White's persistent efforts, however, Roderick Beddow, apparently worn down by the endless vicious infighting, wrote to White at the end of August to announce that he and his firm were withdrawing from the case. White and the NAACP regained the momentum, however, with the announcement that they had recruited the aid of the famous attorney Clarence Darrow. White jubilantly wrote to Willie Roberson:

I have much pleasure in telling you that the N.A.A.C.P. has succeeded in retaining in the Scottsboro Cases the most famous and the greatest criminal lawyer in the United States if not in the world—the great Clarence Darrow known throughout the world as the defender of the defenseless and as the man who always takes the part of the oppressed. . . . I am sure you will be happy to hear this.[57]

Darrow would subsequently be joined by Arthur Garfield Hays of the American Civil Liberties Union, and, during the next several months, White seemed generally upbeat about satisfactory resolution to the question of who would handle the appeals for the Scottsboro cases.

At the same time, White continued his behind-the-scenes campaign of vilification against the Communists. In a letter to the prominent writer Lincoln Steffens, who had been centrally involved in the National Committee for the Defense of Political Prisoners, he wrote:

I have felt up to this time that you were in possession of all the facts regarding the Scottsboro cases. . . . However, on reading the letter just sent out with your name signed to it I see now that you are not in possession of all the facts and that your deep interest in humanity is being used in a fashion that, in our opinion, can only result in increasing the jeopardy in which the boys now are. . . .

The statement quoted in the letter referred to that "Only after the I.L.D. had made Scottsboro a national issue. . . . did the N.A.A.C.P. . . . venture to step timidly into the case" is utterly false. . . . The N.A.A.C.P. was in the case before ever the I.L.D. went into it.[58]

Two days before Christmas, White wrote to Scottsboro defendant Ozie Powell, sending Christmas greetings and a small gift from the NAACP, and announcing the impending visit of Clarence Darrow and Arthur Garfield Hays to Birmingham to review the details of the appeal before the Alabama Supreme Court. Hoping to avoid a repeat of the verbal abuse Roderick Beddow had earlier experienced, White concluded:

We are trying to arrange it so that you can personally talk to the famous men who are to argue your case. We know they will be glad to see you and we hope that if they do come you will receive them with the courtesy which they deserve because of their great interest in you and their hard work on your case.[59]

White could not have had the faintest hint of the dramatic turnabout engineered by the ILD. On December 27, 1931 the following telegram was sent to Clarence Darrow and Arthur Garfield Hays, under the names of all of the Scottsboro defendants except Roy Wright:

We have been getting a lot of letters from the National Association for the Advancement of Colored People telling us that you are going to defend us for the NAACP and we saw the same thing in the newspapers this morning. We have already got the International Labor Defense and jointly employed Mr. George W. Chamlee of Chattanooga and some other lawyers who are helping the ILD. We do not want you to come and fight the ILD and make trouble for Mr. Chamlee just to help the NAACP. If you want to save us and help us get a new trial please help the ILD and Mr. Chamlee. Our parents and kinfolk signed papers yesterday for Mr. Chamlee to file in the Supreme Court that Mr. Chamlee and the ILD are retained by all of us to fight our cases. These papers were filed in the Supreme Court of Alabama today.[60]

The Scottsboro Defense Committee, the ILD, and the League of Struggle for Negro Rights included the telegram in a letter to Walter White, along with the following demands:

We . . . demand that you stop your activities, open and concealed, to disrupt the movement to save the lives of these boys. We demand a public accounting of all Scottsboro funds collected by you. We further demand that you immediately turn over to the Defense Committee all monies falsely and fraudulently collected in the name of the Scottsboro boys.

Shortly afterwards, the following statement appeared in the December 31, 1931 issue of the *Daily Worker* in the names of the Scottsboro defendants:

We. The undersigned. . . . heretofore in April 1931, made a written contract retaining George W. Chamlee, of Chattanooga, and Joseph R. Brodsky and their associates, as our attorneys to make motions for a new trial, and to appeal our cases . . . to the Supreme Court at Montgomery, Alabama, and in May 1931, we made another agreement ratifying the first contract.

Now, in December 1931, we hereby renew our contract with them and authorise them to appear for us in the Supreme Court of the State of Alabama, as we did in our other agreement, and to make the best effort possible, to get our cases reversed and get us a new trial.

We authorized our parents and next of kin to employ Mr. Chamlee and his associates and we hereby ratify all their actions and agreements in reference thereto. . . .[62]

In the face of this totally unexpected onslaught, Clarence Darrow and Arthur Garfield Hays promptly withdrew from the case.[63] Shortly afterwards, the NAACP withdrew, too—at least publicly—but White could not resist a parting shot in a letter to two of the defendants, Willie Roberson and Charlie Weems:

Upon your statements to Mr. Pickens and to me when we visited Kilby Prison, and upon various letters written to us by you and by certain others of the boys, the N.A.A.C.P. at great expense and trouble secured for you the two greatest criminal lawyers in the United States if not in the world, Clarence Darrow and Arthur Garfield Hays. In addition, we engaged the

best law firm in Alabama to aid in your defense. . . . Six trips have been
made from New York to Alabama in your behalf, and all preparations had
been made for the most effective defense possible. . . .

Neither the N.A.A.C.P. nor anybody else can force upon you counsel
which you do not want. I have before me copy of *The Daily Worker* of December 31 in which your name is signed. . . .

You have chosen your counsel and that settles the matter so far as the
N.A.A.C.P. is concerned.

May I say in closing that you and the other boys have vacillated, changing your minds so frequently that it is impossible for any organization or
individual to know just what you do want. May I say to you, however, that
we have no bitterness against you; we understand how perplexed you and
your parents have been. It is unfortunate that you have not been able more
fully to understand all the factors involved in your case.

We wish for you every success under the auspices you have chosen.[64]

White's smooth, self-assured retelling of the Scottsboro case *circa* late 1931
in the December 1931 issue of *Harper's Monthly*, complete with a chronol-
ogy which placed the NAACP at its center from the very beginning, represented the culmination of his intense ideological struggles with the Communists during this phase of the Scottsboro campaign. Published before the
precipitous withdrawal of the NAACP from the appeal before the Alabama
Supreme Court, "The Negro and the Communist" praised the NAACP for
the meticulous investigation and research necessary to mount an effective
legal challenge. Confident that these mechanisms were securely in place,
White turned his attention to the more pressing issues raised by the Scottsboro case.

Acknowledging the global interest in the case without ever once mention-
ing the organizations that had helped to bring that awareness into being,
White presented a graphic account of the facts of the case, complete with
semi-ribald asides and pointed commentary on the peculiarities of the Jim
Crow South. Like the first accounts in the Communist press, White's version
of the story now emphasized the hostility of the white mob that had gathered
at Scottsboro, but it cast the Chattanooga ministers and members of the local
chapter of the NAACP in a heroic light:[65]

A group of Negro ministers and members of the National Association for
the Advancement of Colored People had been stirred to action by the impending danger to the youthful defendants and by the far from groundless
fear that their constitutional rights would be gravely endangered in such an
atmosphere when charged with such an offense.[66]

He also cast Stephen Roddy, the alcoholic and incompetent attorney retained by the Chattanooga ministers, in the best possible terms—"the only

white lawyer in Chattanooga who, so far as they knew, dared face the hostile mob"—even as he acknowledged that "the defense provided the nine boys fell considerably short of perfection."[67] For White, however, the real danger to the Scottsboro Boys was not the inadequacy of the defense, it was the threat presented by the tactics of the Communists:

> Appallingly hostile was the atmosphere already. But that hostility knew no bounds, and the faint chance of getting into the official record of the trial sufficient basis for appeal to a higher court went glimmering when it became known that Judge Hawkins had received a bombastic telegraphic threat from a Communist organization in New York City, the International Labor Defense, which intemperately asserted that the presiding judge "would be held personally responsible unless the nine defendants are immediately released." With sickening rapidity one after another of the boys was found guilty and sentenced to death.[68]

If the Communist Party saw the Scottsboro case as an indictment of the American social, economic, and judicial system, White viewed it as another kind of battleground: "the most notable test of strength to date between those who seek justice for the Negro through American forms of government and those who seek to spread Communist propaganda among American Negroes."[69] In this context, he unleashed his habitual invective against the Communists: they had attempted to corrupt the attorney Stephen Roddy with promises of money and fame, then slandered him with wild rumors and false accusations; they exploited well-meaning liberals and the relatives of the defendants; they had launched a determined attack on the NAACP and "sought vigorously to weaken or destroy confidence in this organization and to injure the reputation it had built up over a period of two decades"[70]; they had stirred up trouble in Camp Hill, Alabama in July 1931 when a meeting of black sharecroppers and tenant farmers turned into a shoot-out that left a sheriff wounded, one farmer dead and five wounded, a church burned to the ground, and dozens arrested; and they had been involved in eviction struggles in Chicago that left two African Americans dead and others wounded. No *white* Communists were injured in these episodes, White was quick to point out, the surest sign that they were cynically manipulating blacks for their own nefarious purposes.

White freely acknowledged the dire social, economic, and political conditions many African Americans faced in Depression America, buttressing his claims with hard facts and statistics. And he noted as well the stirrings of consciousness that had occurred within African American life since the end of World War I, itself a product of seismic shifts in global conditions:

> More and more the Negro has realized that the problem he encounters in the United States is only a part of the whole structure of race prejudice which the modern age of imperialism has created. The stirrings of thought

and unrest in Africa, Asia, and the Far East of black and brown and yellow races have not been without their counterpart in the United States. The War shattered whatever notion of the white man's infallibility and invulnerability the Negro and other colored races unquestionably held prior to 1914.[71]

The existence of this "embittered Negro world," White cautioned his readers, certainly presented fertile soil for Communist agitation and influence, but, he assured them, Communist leaders were too shortsighted to take advantage of these conditions: "Had they been more intelligent, honest, and truthful there is no way of estimating how deeply they might have penetrated into Negro life and consciousness." But their "bungling and dangerous tactics in the Scottsboro cases" had shifted the pendulum away from them.

Against the bogeyman of the Communist threat White pitted the historical loyalty of African Americans to American values and institutions. Still, he warned:

> There is but one effective and intelligent way in which to counteract Communist efforts at proselyting [sic] among American Negroes, and that method is drastic revision of the almost chronic American indifference to the Negro's plight. Give him jobs, decent living conditions, and homes. Assure him of justice in the courts and protection of life and property in Mississippi as well as in New York. . . . In brief, the only antidote to the spread among American Negroes of revolutionary doctrine is even-handed justice.[72]

Wily, ruthless, and unforgiving of those he regarded as his enemies, White skillfully sought to turn the presence of the Communists in the Scottsboro case to the advantage of the agenda of the NAACP; and he pressed his claims upon the attention of liberal readers through the pages of magazines like *Harper's Monthly*, *The New Republic*, *The Nation*, and *Christian Century* in the United States; the *New Leader* in England; and *Les Cahiers des Droits de l'Homme*, where Magdeleine Paz's article, "L'Affaire de Scottsboro," essentially followed White's version of events.[73]

In the short run, Walter White and the NAACP lost the tactical and political skirmishes with the ILD and the Communists over the conduct of the Scottsboro defense, but White's relentless counterattack took its toll on the Communists, too, and, over time, the NAACP version of events—especially its public doubts about Communist motives, its attacks on the tactics and ultimate goals of the ILD and the Communist Party, and its persistent queries about the financial accountability of the ILD—assumed a central place in the public discourse about the Scottsboro case.

Ironically, the virulently anti-communist views of the NAACP sometimes significantly overlapped with the general outlook of the southern press, public officials, and intellectual apologists. In Alabama, the dominant point of view was eloquently represented by Grover Hall, editor of the *Montgomery Advertiser*, the recipient of a 1928 Pulitzer Prize for his probing exposés of the influence of the Ku Klux Klan on Alabama politics and culture, a liberal—in southern terms—whose journalistic sensibilities had been shaped by the iconoclasm of his friend and idol H. L. Mencken.

From the very beginning of the arrests at Scottsboro, Hall hailed the restraint of the sheriff and the citizens in the region. This was, in fact, the first line of defense of the state of Alabama, and the South in general: the expected ritual had *not* occurred. There had been no mob violence, no lynchings. The reason a large crowd had gathered at Scottsboro for the trials was because it was the first Monday of the month, a traditional market day for farmers from north-central Alabama. There was no intimidation of the jurors. Hall not only praised the restraint of his fellow Alabamians, he also directed the attention of his readers to the confessions of several of the Scottsboro defendants and their previous police records, and to the testimony of Victoria Price and Ruby Bates. The Scottsboro Boys were guilty, Hall concluded, and their trials were fair.[74] The only real trouble had been caused by "outside agitators" in the form of Communists and their supporters, whose activities, Hall feared, might create the conditions for the return of the Ku Klux Klan, which had only recently been driven out of Alabama political life. Hall remained steadfast in his beliefs until 1937, when he finally announced an abrupt and eloquent about-face after several meetings with Allan Chalmers of the Scottsboro Defense Committee.[75]

In the early stages of the Scottsboro campaign, Hall's views were generally echoed by J. Glenn Jordan, city editor of the *Huntsville Times*, in a controversial pamphlet offered for sale at the 1933 Decatur trials before Judge Horton impounded it and forbade its sale. "I am writing this book," Jordan declared,

> . . . not because of any hope of great financial gain but because I want to bring home to the people of America the gravity of the alarming growth of the "Red" movement in the United States and the means they are using to stir up racial misunderstanding in this country; because I want to tell them that unless something is done and done quickly to stop the Russian born propaganda America will soon be facing a situation just as alarming as she did in those grave days of 1917–18.
>
> I want to point out that the people of my glorious South have been grossly misjudged and accused and that such propaganda as has been circulated on the Scottsboro case is doing more to stir up racial strife than any other one thing.
>
> I want to add my defense to the people of Jackson County, Ala., a law-abiding, peaceful people, who are smarting under the unjust accusations. I hope to show the people of the North, the East and the West, that we of the South are not barbarians and DO give the negro race a square deal.[76]

Jordan's claims to authority were based on his assertion that he was "one of two accredited newspaper correspondents who were assigned to cover the Scottsboro case from beginning to end," and he insisted "every statement in this book is based on the records of the case."

The cornerstone of Jordan's "defense" of the state of Alabama, like Grover Hall's argument, rested upon his insistence, as an eyewitness, that there was never any threat of mob violence when the Scottsboro Boys were transported to town. In his telling, the people gathered there became visitors:

> It seemed to me that the entire population of the little county seat, aug-mented by hundreds of visitors, surrounded the two-story dilapidated jail. . . . Only a slow mumbling of hundreds of voices in casual conversation came from the outside. There were no loud outbursts of any nature.[77]

And he corroborated his assertion by recounting the following exchange with the sheriff:

> "Sheriff?" I questioned him, "I noticed several hundred people outside as I came in. Do you look for any attempt at mob violence?"
>
> "None at all," he replied firmly. "Those people will let the law take its course. You are in Jackson County where they believe in a square deal for every one. I am not worried. While this is a case that will raise a storm of in-dignation in the heart of every red-blooded man and woman, I do not fear any violence on the part of my people here in Jackson County," he added.
>
> "But sheriff," I continued. "As I came in I heard that you had asked that troops be sent to Scottsboro tonight. Why is that?"
>
> "There is a great crowd here now," he explained. "And it will grow larger as the news spreads. There will be many people here from other counties. I feel that it is my duty as an officer to take every precaution to protect my prisoners. There is no more sentiment here than naturally arises from such a charge. It was on the grounds of the offense and not from any voice of feeling that I asked for military assistance."
>
> And so in a few words he dismissed any idea that the negroes might be lynched. There had been numerous other incidents, both in the North and the South, where troops had been called out to maintain order. Sheriff Wann was only doing his duty as a sworn officer of the law.[78]

Having thus dispensed with the lurid and clearly patently false accounts of mobs gathered at Scottsboro, Jordan offered a succinct account of the events leading to the Scottsboro trials, including the somewhat rambling exchange among Judge Hawkins and attorneys Stephen Roddy and Milo Moody about who would represent the Scottsboro Boys. In a similar fashion, Jordan's "evi-dence" against the defendants drew heavily—and selectively—from the more than seven hundred pages of court transcripts of the case. Victoria Price's lurid, detailed testimony occupied a central place in Jordan's text, as did the damning testimony of Clarence Norris, who testified that all of the defendants had assaulted the two women, except him; and Roy Wright, who testified that nine blacks had had intercourse with the girls, but not the ones from Chattanooga.

In his discussion of the verdict of the first trial, and its aftermath, Jordan took the opportunity to rebut the widespread accusation that the juries were influenced by the visible and audible expression of passions aroused by the case:

> A mild cheer and applause followed the reading of the verdict, but it was promptly hushed by a sharp warning from the bench. It was next to impossible for the second jury to have heard the reading of the verdict and highly improbable that it was conscious of the cheering. People in the court house yard declared they could hear nothing from the court room.[79]

He also pointed to the "avalanche of messages which have deluged public officials of Alabama, for nearly a year. . . , many of them containing sinister threats," and violent protests in Germany and New York City—laying most of these efforts squarely at the feet of the ILD. Jordan briefly reviewed the circumstances under which Clarence Darrow and Arthur Garfield Hays withdrew from the Scottsboro case, then turned his attention to what he referred to as the "Dundee Episode"—the incident revolving around the statement Ruby Bates wrote repudiating her courtroom testimony—linking the attorney for the Scottsboro Boys, George Chamlee, by implication to the byzantine plot to solicit Ruby Bates's "confession."

All of Jordan's "evidence" was delivered at a breathless pace, and with at least the pretence of journalistic objectivity, and all of it pointed to its foregone conclusion:

> And this brings the story of the Infamous Scottsboro case to a close as far as this little book is concerned. The mother of Roy and Andy Wright has sailed for Europe for a three months "lecture" tour, sponsored by the International Labor Defense League. It is believed the Communists will make a last grandstand play in behalf of the negroes in order to win new members and to enrich their coffers, and then Alabama and the world will finally see justice claim its own—will see the negroes receive their just deserts—death in the electric chair.[80]

To insure that its intention was not lost upon its readers, however, the publisher's appended the following note:

> This little book is being offered to the reading public as an answer to the malicious and libelous propaganda of certain Communist organizations which are attempting to capitalize on the conviction of the seven negro fiends who were unquestionably guilty of the crime with which they were charged.
>
> We hope you will enjoy reading it as much as we have in publishing it.[81]

Southern intellectuals also rose up in defense of the South, most notably Frank L. Owsley, who cast the Scottsboro case as the latest act in a historical series of epic battles between the North and the South that had been waged since the Missouri Compromise debates began in 1819.[82] A professor of Southern and Civil War History at Vanderbilt University and one of the twelve contributors to the powerful manifesto that defined the Agrarian movement in 1930, *I'll Take My Stand: The South and the Agrarian Tradition*,[83] Owsley revisited the issue he had explored at length in that volume: the "irrepressible conflict" between the North and the South.[84]

Arguing that "the mistake most often repeated has been the interference of the North—usually the Northeast—with the relationship between the whites and blacks of the South,"[85] Owsley sought to place this pattern in the widest possible historical framework:

> Twice, indeed, it has developed into war upon the South: the abolition crusade of 1819–1865 (including the Civil War); and the period of reconstruction. In the present epoch the interference is again rapidly assuming the proportions of a war upon the South. The Scottsboro and Dadeville cases, Judge James Lowell's action in the case of the Negro refugee murderer from Virginia, and the wide-spread agitation involved, are some manifestations of the third crusade.[86]

The northern writers, artists, and intellectuals who were at the forefront of the defense of the Scottsboro Boys, Owsley argued, "are intellectual children and grandchildren of the abolitionists and reconstructionists," and he committed himself to the task of exposing their historical antecedents: "Indeed, if I can demonstrate the analogy both of technique and motive between the abolition crusade, reconstruction, and the present agitation for 'justice to the Negro' as exemplified in the Scottsboro affair, I shall consider that I have performed a patriotic duty. For I am exceedingly anxious and seriously alarmed over the present agitation."[87]

Against the backdrop of the ceaseless sectional rivalry between the North and the South, Owsley maintained:

> The industrialists, carefully coached by their lawyers and statesmen and "intellectual" aides, realized the bad strategy of waging a frank struggle for sectional power; they must pitch the struggle upon a moral plane, else many of the intelligentsia and the good people generally might become squeamish and refuse to fight. Shibboleths and moral catchwords must be furnished. It was therefore found convenient to attack slavery as an evil and the slaveholder as a criminal, in fact to impugn the morality of the South, in order to create opinion in the East and North in favour of the industrialists' plan of Southern restriction.[88]

Deeply embedded in Owsley's account was the paternalistic view of relationships between blacks and whites that was the cornerstone of the myth of southern plantation life. The immediate consequence of the northern assault upon southern institutions, he argued, was the disruption of previously harmonious race relations. The sudden emancipation of blacks "had the effect of bringing about the ruin, for many years, of both himself and his former master;" and during Reconstruction the northern emissaries of the Radical Republicans "preached to the naïve and childish blacks incendiary doctrines calculated to rouse the wildest passions of revenge against their former masters."[89]

The collapse of Reconstruction and its aftermath, however, carried with it a stern warning for the future, Owsley emphasized:

> *The Negro should have learned then that in the end the relationship good or bad between himself and the white race of the South must be settled between them and that no outside power could dictate permanently the terms of this relationship.*[90]

The subsequent pattern of relationships between blacks and whites in the post-Reconstruction South has been shaped by this history, Owsley continued:

> The white race disciplined him severely for his conduct during reconstruction. At length, however, with outside interference largely removed, old friendships were renewed between the races, new ones were formed until something of the old affection which had existed between the black man and the white man returned. His condition gradually improved; though he has never been given back the ballot nor allowed to sit on juries. The experience of reconstruction was too bitter to be soon forgotten.[91]

The journalists, poets, novelists, and "intelligentsia" now spearheading the "third crusade" against the South, Owsley charged, were simply following the well-worn path of their predecessors: "The method is familiar: holding the South up to ridicule as backward, ignorant, unprogressive. . . . giving wide currency to race conflicts and lynchings in the South, while ignoring such difficulties in the North as the Chicago and St. Louis race riots."[92]

To Owsley's jaded eyes, it was the "familiar spectacle of the intelligentsia taking the lead in this new thrust between the white and black man of the South":

> Such diatribes as that of Mrs. Mary Heaton Vorse in the *New Republic* on the Scottsboro case sound alarmingly like the pages out of a Northern journal during the days of reconstruction; and one might well suspect, in the light of history, that the public mind in the North is being prepared for the

identical program. The robes of morality are having the dust and wrinkles shaken off for the intelligentsia and good people to march forth in. One feels assured that industrialism has found the agrarian South once again an obstacle in the way of its control of the federal government.[93]

For Owsley—and, indeed, for his embattled Agrarian colleagues—the Communists were a subset of the "industrialists" who had historically tried to subjugate the South, but the program of the Communist Party—as he gleaned it from his close reading of *The Communist Position on the Negro Question*— was particularly inflammatory and insidious not because it advocated social- ism, but because it promised black people land:

> It urges the Negro to rise and seize the lands of his oppressors and divide them out. It promises to confiscate and give him these lands should the Communist Party succeed in gaining control of the Federal Govern- ment. It even promises to create a Negro state out of the Southern black belt and grant the Negro autonomy or allow him to secede from the union. . . .[94]

This is the political dynamic fueling the brouhaha around the Scottsboro case, Owsley maintained:

> These young Negroes, while it is said they are not Communists them- selves, had been embittered and inflamed against the Southern whites be- fore the alleged attack upon the two white women. Certainly since the beginning of the case the Communists have made every possible use of it to incite these and other Negroes to do violence against the whites. They say in their Southern Program that the Scottsboro case is their most strate- gic point in the world at the present time. They call attention to the wide- spread support given them in the North by people who are not Commu- nists—in fact, who are thoroughly unaware that they are helping a Communist program. At the present time the "Scottsboro boys," as the Communists love to call them, are holding levee in the Birmingham jail. Hundreds of Negroes, not Communists, of course, but Negroes who have been aroused by the incendiary literature meant to set the blacks to fight- ing the whites, visit the jail.[95]

Like Walter White—albeit for significantly different reasons—Owsley ar- gued: "once again, the unfortunate black is being used as a pawn"; the Com- munists "would prefer that the 'Scottsboro boys' be executed by the State of Alabama. . . . This would create another group of martyrs, another period of violent propaganda, in which the Southern people, their courts, their govern- ments, would be held up to ridicule and Northern opinion further prepared for a Southern program; while more fuel would be added to the race antago- nisms in the South."[96]

Like Grover Hall, Owsley darkly predicted that continued "outside agitation" by northerners, particularly Communists, could only lead to the resurgence of the Ku Klux Klan and other extralegal means of "disciplining" southern blacks:

> Such cases as that of the "Scottsboro boys" would be tried in courts whose decrees could not be appealed to a Federal tribunal. In other words, the outside interference . . . can result in nothing but organizations like the Ku Klux Klan and in violent retaliation against the Negroes—themselves often innocent. Such interference makes justice difficult, not only by creating friction between the races, but by arousing anger against those who interfere: Lawyer Leibowitz [who represented the Scottsboro boys at their second trial in Decatur in 1933] and those who sent him were more on trial at Decatur than the Negro boys. The boys might have been cleared by a native lawyer on the second trial.[97]

Owsley conveniently neglected to mention that a "native lawyer," the Chattanooga attorney hired by the ILD to handle the Scottsboro Boys' defense, had been involved in the case from almost the beginning—with no apparent success. Still, his article represented one of the most intellectually sophisticated apologies for the South to date. More dispassionate, and much less personally invested than many of the southern journalists who covered the case ("I am not here concerned with the guilt or innocence of the Negroes—although I wish justice done," he wrote[98]) Owsley, through his sober reflections on the broad pattern of relationships between the North and the South, added the veneer of scholarly authority to the striking admixture of defiance and defensiveness that constituted the widespread southern reaction to public scrutiny about the Scottsboro case. In the final analysis, though, his perspective was not that significantly different from the views popularized by the mainstream southern press. In all cases, the essential argument was that the South, left to its own devices, could be counted upon to resolve and properly regulate the relationships between blacks and whites, that this was essentially an internal matter not subject to outside interference, particularly the sinister schemes of the Communists—a point of view that flew in the face of the glaring fact that, had the outpouring of public outrage about the Scottsboro Boys not occurred, it is likely that eight of them would have been executed on July 10, 1931.

The southern fixation upon the Communist menace reached a peak with Files Crenshaw, Jr. and Kenneth A. Miller's *Scottsboro: The Firebrand of Communism*, a voluminous work that not only drew extensively upon the verbatim testimony of the trials, but also introduced carefully selected photographs of the Scottsboro defendants to make its case: one of a sullen, menacing Haywood Patterson, with a suggestive gaping hole in the crotch of his trousers, the other of Andy Wright, Charlie Weems, and Clarence Norris in "urban" clothes, financed, no doubt, by their liberal and Communist supporters.[99]

Crenshaw and Miller began with the usual disclaimers: "The sole purpose in publishing this book is to refute the slur that has been cast upon the State of Alabama and the South and it is hoped that anyone will, in a spirit of sportsmanship and fair play, go to the actual testimony."[100] In this instance, the "actual testimony"—like that compiled by J. Glenn Jordan in an earlier stage of the history of the case—consisted of carefully culled and edited excerpts from the official transcripts of the case.

Through hundreds of brain-numbing pages, Crenshaw and Miller revisited the testimony of the trials at Scottsboro and Decatur, the second trial of Clarence Norris, and the third and fourth trial of Haywood Patterson, always giving primacy of place to the lurid accounts of Victoria Price, and generally keeping editorial commentary to a minimum. Only someone already familiar with the details of the trials would notice that the testimony of Clarence Norris—who provided, and later recanted, the most damning testimony against his codefendants—is missing. Gaps such as these indicate the method by which they slowly but systematically develop their argument. Through innuendo, insinuations, and non sequiturs, Crenshaw and Miller accumulate their damning evidence against the Communist conspiracy. Thus, in their brief account of the ILD's entry into the case: "The authors pause here to reveal, through its own literature, the deception employed by the I.L.D. to trick the unsuspecting negro into its organization and, at the same time, cleverly conceal its true identity."[101]

Later in the book the authors challenge the Communists' designation of the boys as "workers," pointing out that "nowhere in the record is there anything to indicate that any of them were even remotely connected with Southern industry," before moving on to their more central point:

> Since the first slave was brought into this country by New England traders the white race has insisted that the negro respect the white woman, regardless of her position in society, and when he violates the person of a white woman he has committed "the unpardonable sin" in the eyes of the Southern white.
>
> In the trials of the Scottsboro negroes their counsel has attempted to minimize the seriousness of the crime with which they are charged by assailing the characters of the two women involved. The characters of these Huntsville girls, whatever they might be, are not the concern of the authors, or of the courts. They could have belonged to the most select social set in this beautiful North Alabama city and have been traveling the highway in a limousine instead of riding the freight.[102]

Their example is far from hypothetical, as the authors proceed to describe in graphic detail the murderous assault by a black man upon three socially prominent young women from Birmingham in August of the same year that Scottsboro occurred. In this manner the Scottsboro Boys were indicted by implication in this event. But Crenshaw and Miller also seize upon it to dramatize how "law and order had triumphed in Alabama to prevent wholesale bloodshed":

> negro organizations in Birmingham offered every assistance at their command in the search for the slayer, posted rewards, adopted resolutions deploring the shooting of the girls and expressed the belief that the murderer was no product of Alabama and "possibly came to this state for the purpose of making trouble in the furtherance of propaganda to which both races in the South are opposed."[103]

This scenario of a cooperative local black community that eagerly joined in the search for an "outside agitator" is skillfully linked to a gesture of leniency by white authority: after the assailant was identified and apprehended, "he was tried and sentenced to die, but Governor Miller was never completely convinced he was the perpetrator of the crime and commuted his death sentence to life imprisonment."[104]

This is the kind of evenhanded justice the Scottsboro Boys could expect, too, Crenshaw and Miller imply. In their brief account of the entry of the NAACP, Clarence Darrow, and Arthur Garfield Hays into the case — and their withdrawal — Crenshaw and Miller seem to give the NAACP backhanded praise for respecting southern sensitivities and prerogatives in the case:

> Darrow and Hays had been sent to Birmingham by the National Association for the Advancement of Colored People to represent the condemned negroes. However, they revealed it was their plan to remain in the background and permit some prominent Southern barrister to take the lead in the defense.[105]

When they deem it necessary, Crenshaw and Miller intervene to defend the character of the Alabama public officials who have been maligned, threatened, and libeled by the Communists, as they do for Governor B. M. Miller, who "from his experiences on the bench had developed a sympathy for mankind that few men with his judicial and executive background possess" and "came from staunch Presbyterian stock;"[106] and Lieutenant Governor Thomas Knight, who "was reared at Greensboro, Alabama, and is a descendant of one of the oldest and most prominent families in that part of the State. . . . His grandfather had been a captain in the Confederate army. . . . Knight is a member of the National Conference of Jews and Christians, which should dispose of any charge of his being a bigot."[107]

They also take aim at Alabama's *bete noire*, Attorney Samuel Leibowitz: "Leibowitz' record as a criminal lawyer in New York stamped him as a wholesale defender of alleged murderers and disclosed him as a barrister not at all averse to publicity, and perfectly willing to take the weaker side of a case if the laurels of a possible victory were sufficiently promising."[108]

Figure 1.6 Andy Wright, Charlie Weems, and Clarence Norris. Source: Files Crenshaw, Jr. and Kenneth A. Miller, *Scottsboro: The Firebrand of Communism* (Montgomery, AL: Brown Printing Company, 1936).

To sum up, write Crenshaw and Miller at the end of their exhaustive proceedings,

> Having read the record of the Scottsboro case, this question at once presents itself: "Are the Communists and their sympathizers even remotely interested in securing justice for the nine negro defendants, or are they using them merely as tools to create animosity between the white and black races.?"[109]

For these authors the most egregious manifestations of the Communist conspiracy are the "more than 15,000 telegrams, cablegrams, letters and cards heaping ridicule and condemnation upon Alabama"[110]—a breach of decorum and etiquette that places the Communists beyond the pale of civilized behavior, and activity that places the state of Alabama in a position analogous to that of Victoria Price and Ruby Bates:

> The red hordes of Communism that have swarmed into Alabama since that rattling freight deposited its human cargo in Paint Rock are intent on a much greater crime than any that might have been committed in the black gondola. They would ravish the laws and social standards that condone racial discrimination and bury them in the blackened ruins of revolution.[111]

Fittingly, Crenshaw and Miller conclude their indictment of the Communists, and their defense of the state of Alabama, by returning full circle to the words of Grover Hall, the editor of the *Montgomery Advertiser*:

> Upon two recent famous occasions county officials in Alabama have acted in the best tradition of civilized government to protect "niggers" charged with the highest of all crimes known to the Deep South.
>
> First, Scottsboro; second, Huntsville.
>
> In the first instance a gallant and intelligent sheriff saved the lives of a band of men charged with the First Crime from death—they were men who had never lived in Jackson county, they were charged with assault upon women who had never lived in that hill county. They could have been lynched in High Jackson. But High Jackson has lived to see itself libeled and humiliated on every continent nevertheless. It has seldom been praised by a commentator outside of Alabama for its restraint in the face of torrential passion.[112]

The "torrential passion" unleashed by the Scottsboro case generated an extraordinary range of rhetorics—all of which had direct bearing upon the ways in which the images of the Scottsboro Boys were crafted, and circulated; that is to say, they constituted the range of possibilities within which the Scottsboro defendants were imagined, the grounds upon which the narratives about them would be constructed. In the early stages of the construction of Scottsboro Narratives, the version constructed by the Communist Party quickly gained ascendancy.

CHAPTER TWO

"SCOTTSBORO, TOO"

The Writer as Witness

> I said:
> Now will the poets sing,—
> Their cries go thundering
> Like blood and tears
> Into the nation's ears,
> Like lightening dart
> Into the nation's heart.
> Against disease and death and all things fell,
> And war,
> Their strophes rise and swell
> To jar
> The foe smug in his citadel.
>
> Remembering their sharp and pretty
> Tunes for Sacco and Vanzetti,
> I said:
> Here too's a cause divinely spun
> For those whose eyes are on the sun,
> Here in epitome
> Is all disgrace
> And epic wrong,
> Like wine to brace
> The minstrel heart, and blare it into song.
>
> Surely, I said,
> Now will the poets sing.
>   But they have raised no cry.
>   I wonder why.
>
> —Countee Cullen, "Scottsboro, Too, Is Worth Its Song"[1]

CONTRARY TO THE ASSERTION of Countee Cullen's poem, American poets were among the first to respond to the appeals of the defenders of the Scottsboro

Boys, often effectively capturing the immediacy of the moment in ways that simply were not possible for writers in other genres. Cullen's labored breathing in "Scottsboro, Too," on the other hand, has a johnny-come-lately feel to it.

First published in the official newspaper of the American Communist Party, *The Daily Worker*, in October 1933,[2] with the subtitle "a poem to American poets," Cullen's poem is a curious and disingenuous public performance—particularly since it appeared several years after the Scottsboro case had settled into the dreary pattern of litigation and recrimination that would characterize it for the rest of the decade. The disingenuousness of the poem rests in the posture of aggrieved innocence that pits the speaker's apparent passionate concern about the legal and moral travesty of Scottsboro against the presumed indifference of the community of American poets. Given the publication within which the poem appears, it seems that Cullen's criticism is directed at the behavior of "bourgeois" American poets, but there is also the very strong implication that the "I" is distinctly African American, while "They" are white; hence, the charge that the subject of nine young, poor black men is not deemed sufficiently worthy for (white) American poets. The speaker recalls how ""they" had composed "sharp and pretty / Tunes for Sacco and Vanzetti," the political anarchists convicted of robbery and murder by circumstantial evidence in 1921 and executed in 1927—in sharp contrast to their apparent neglect of the Scottsboro case.

Ironically, the speaker in "Scottsboro, Too" does not reveal that the poet Countee Cullen was among the many artists and intellectuals for whom the Sacco and Vanzetti case had become a *cause célèbre* in the 1920s; many contemporary readers of Cullen would have known that he had excoriated Sacco and Vanzetti's judges in the stately, elevated diction of his sonnet "Not Sacco and Vanzetti":

> These men who do not die, but send to death,
> These iron men who mercy cannot bend
> Beyond the lettered law; what when their breath
> Shall suddenly and naturally end?
> What shall their final retribution be,
> What bloody silver then shall pay the tolls
> Exacted for this legal infamy
> When death indicts their stark immortal souls?[3]

In fact, the speakers of "Not Sacco and Vanzetti" and "Scottsboro, Too" seem closely linked in bearing and diction, suggestive of the temperamental and aesthetic difficulties Cullen may have encountered in trying to write poems of direct political engagement. At any rate, the claim in "Scottsboro, Too" that (white) American poets had failed their moral, political, and poetic responsibilities by not raising a thunderous public outcry about the case, is as unpersuasive as poetry as it was untrue in fact.

In spite of the aggrieved posture of the poem's speaker, Countee Cullen himself had followed the details of the Scottsboro case with great interest almost from the very beginning. In early July 1931, Cullen sent a cable from Paris (where he had vacationed every summer since 1926)[4] asking his friend Walter White, Executive Secretary of the NAACP, for the latest news about the Scottsboro defendants. Later that month he wrote White:

> Thanks much for your telegram explaining the Scottsboro case. There had been no news here until your telegram came. It may not please you because of your troubles with the American communists to know that I wanted the news for use by some communist friends here. They have been the only ones, as far as I know, who have shown an active interest in the case here in France. I hope you don't mind my having given your communication to them.[5]

By the summer of 1931 Cullen was certainly sufficiently aware of the political dynamics of the Scottsboro case to know that his friendly association with French Communists would not please Walter White; in fact, there seemed to be at least a hint of defiance in Cullen's tone towards White. Furthermore, Cullen would have known and undoubtedly identified with the signatories of a March 1932 petition to President Herbert Hoover from Americans then resident in Paris:

> We . . . after having attentively followed the Scottsboro case from its beginning to the unjust conviction of the eight Negro youths, feel most profoundly constrained to address to you this solemn declaration of our opinion. At the same time, Mr. President, in the name of JUSTICE, HUMANITY and FAIR PLAY we humbly beseech you to use the influence of your high office and make intervention on behalf of these eight American children.
>
> This unfortunate incident offers the occasion for the eyes of the whole world to be centered on our beloved country. The nations wonder how the Federal Government of our great United States—the land of liberty and justice—can stand idly by and tolerate legalized lynching of mere children.
>
> It is extremely perplexing to the civilized mind to conceive how such a horrible crime, the murdering of eight human beings, can be countenanced by a people who profess the religion of our lord and savior JESUS CHRIST. And for what have these youths been convicted?? For an ALLEGED "rape" of two females of doubtful character in company with seven vagabonds in a box car.
>
> The disrespect for law caused by prohibition, the reign of bandits and the Lynch law cause us to hang our heads in shame when other nationals say that America, our America—the land of WASHINGTON and LINCOLN—is actually the most lawless and barbarous country on earth. . . .[6]

By the early 1930s Countee Cullen—like many American writers and intellectuals—had drifted leftward. In 1932 he publicly declared: "The Communist Party alone is working to educate and organize the classes dispossessed by the present system,"[7] and he joined his friend Langston Hughes in formally endorsing the Communist Party candidates for President and Vice President of the United States, William Z. Foster and James W. Ford. The following year he publicly endorsed a national two-day anti-lynching conference in Baltimore, sponsored by the League of Struggle for Negro Rights; and in a letter to the League's newspaper *Negro Liberator* he praised African American Communist Party leader William Patterson and Langston Hughes for their "effort to awaken the country at large to a realization of the deadliness of this terror."[8]

Given his new-found political commitments, Cullen would have been well aware that the Scottsboro case had indeed captured the attention of the American public, including the community of American poets. By the early 1930s poems about Scottsboro routinely appeared in the NAACP's *Crisis* magazine, the Urban League's *Opportunity*, the Communist Party's *New Masses* and *The Daily Worker*, other Left magazines like the *Anvil* and *Left Front*, and the daily press as well. How, then, to account for the dyspeptic and accusatorial attitude of "Scottsboro, Too, Is Worth Its Song"?

"The most widely celebrated and probably the most famous black writer in America during the heyday of the period that has come to be known as the 'Harlem Renaissance,'"[9] Cullen was best known for his skillful mastery of traditional poetic forms such as the ballad, Shakespearean and Petrarchan sonnets, and Spenserian stanzas. He was also deeply committed to the view that poetry was, or should be, a sublime experience. As he remarked in a 1926 review of Langston Hughes's first collection of poetry, *The Weary Blues*, he had serious doubts about whether "all subjects and forms [are] proper for poetic consideration."[10] In this instance, he was speaking about jazz—which he regarded as an insufficient resource for serious poetry—but, in a broader sense, he was articulating the aesthetic principles that were the bases of his poetic practices. These same principles are at work in "Scottsboro, Too," and they offer insights into the dilemma the poem addresses.

According to the speaker, Scottsboro is a "divinely spun" cause, an "epic wrong" that should "brace / The minstrel heart, and blare it into song." The structure of the poem, however, and its elevated diction has the opposite effect—it distances readers from the event it invokes. Not known for his poetic forays into the world of political events, Cullen falters here. Perhaps "Scottsboro, Too" is better understood as his own belated attempt to enlist the muse of poetry on behalf of a cause in which he believed—and as a doleful self-criticism of his inability to rise to the occasion.

On the other hand, Langston Hughes—Cullen's contemporary and sometimes intellectual jousting partner—proved himself to be much more adept at fashioning effective art out of the plight of the Scottsboro Boys. Deeply emotionally and politically invested in the case from the very beginning, Hughes devoted considerable attention to it in essays, poetry, and drama. As Hughes's biographer Arnold Rampersad succinctly observes: "In the fall of 1931, the driving public force in Hughes's move to the left was certainly the Scottsboro controversy."[11] His earliest response to the case, "Scottsboro," appeared in the Urban League's *Opportunity* magazine in December 1932:

8 BLACK BOYS IN A SOUTHERN JAIL.
WORLD TURN PALE!

8 black boys And one white lie.
Is it much to die?

Is it much to die when immortal feet
March with you down Time's street,
When beyond steel bars sound the deathless drums
Like a mighty heart-beat as They come?

Who comes?

Christ,
Who fought alone.

John Brown.

That mad mob
That tore the Bastille down
Stone by stone.

Moses.

Jeanne d'Arc.

Dessalines.
Nat Turner.

Fighters for the free.

Lenin with the flag blood red.

(Not dead! Not dead!
None of these is dead.)

Gandhi.

Sandino.

Evangelista, too,
To walk with you—

8 BLACK BOYS IN A SOUTHERN JAIL.
WORLD, TURN PALE![12]

The opening lines function as a headline, an invocation, and a refrain, but there is a sharp contrast between the terse first two stanzas of the poem and the third—where the poem seems to shift its attention from the question of how the world should behave in the face of this legal travesty to an exhortation to the boys themselves about how they should try to face their impending execution. The "immortal feet" that will march with the Scottsboro Boys "down Time's Street" constitute a role call (albeit a very eclectic one) of inspirational martyrs, rebels, and revolutionaries. There is little indication in the poem, however, that the world can do anything more than bear witness to the martyrdom of the Scottsboro Boys.

During the fall of 1931, Hughes also wrote *Scottsboro Limited: A One-Act Play*,[13] which was first published in *New Masses* and reappeared the following year as *Scottsboro Limited: Four Poems and a Play in Verse*, with illustrations by the talented white artist Prentiss Taylor.[14]

Scottsboro coincided with Hughes's first poetry reading tour of the South, an eight-month saga that began in October 1931, and, from the very beginning, the case functioned as a touchstone for his rapidly evolving social and political attitudes.

Hughes's first sustained encounter with "the ethics of living Jim Crow," as Richard Wright would define southern racial practices later during the decade, occurred at Hampton Institute, Virginia, in early November 1931. On the weekend Hughes arrived he learned that Juliette Dericotte, the beloved Dean of Women at Fisk University, had been seriously injured in an automobile accident in Georgia. Much as in the legendary account of the death of the blues performer Bessie Smith, Juliette Dericotte was denied emergency treatment at a white hospital and died before she could be transported to a hospital for blacks, many miles away. Hughes also heard of the death of a former Hampton athlete, a coach at Alabama Normal College for Negroes, who had been beaten to death by a white mob when he was on his way to watch his team play its first game of the season. He had mistakenly parked his car in the parking lot reserved for whites.[15] Hughes was horrified and outraged by these reports and immediately agreed to speak at a student protest meeting:

> But no such meeting was held at Hampton. The elderly white and Negro heads of the Institute would not permit the students to hold a protest meeting. They—and I—were told at a student committee gathering, "That is not Hampton's way. We educate, not protest."

In an essay published in *The Crisis* three years later, "Cowards from the Colleges," Hughes was still fuming about what he regarded as a conspiracy by many historically black colleges and universities:

> Many of our institutions apparently are not trying to make men and women of their students at all—they are doing their best to produce spineless Uncle Toms, uninformed, and full of mental and moral evasions.[16]

For Hughes, the clearest evidence of the moral and political bankruptcy of these institutions—and the basis of his scathing indictment of the "mid-Victorian attitude in manners and morals" that governed them—was their studied indifference to the plight of the Scottsboro defendants:

> I was amazed to find at many Negro schools and colleges a year after the arrest and conviction of the Scottsboro boys that a great many teachers and students knew nothing of it, or if they did the official attitude would be, "Why bring that up?" I asked at Tuskegee, only a few hours from Scottsboro, who from there had been to the trial. Not a soul had been so far as I could discover. And with demonstrations in every capital in the civilized world for the freedom of the Scottsboro boys, so far as I know not one Alabama Negro school until now has held even a protest meeting.[17]

Some of Hughes's earliest poems and public statements about the Scottsboro case were a direct outgrowth of his first southern tour. From the University of North Carolina at Chapel Hill, one of the rare white institutions to arrange a booking for him, Hughes was contacted by the two student editors of the magazine *Contempo: A Review of Books and Personalities*, Anthony J. Buttitta and Milton Abernethy, specifically soliciting something from him about Scottsboro. Hughes responded with an essay, "Southern Gentlemen, White Prostitutes, Mill Owners, and Negroes," and a poem, "Christ in Alabama,"[18]—two highly provocative works which Buttitta and Abernethy published in the December 1931 issue of *Contempo* to coincide with Hughes's campus appearance.

"Southern Gentlemen, White Prostitutes" is an arch, mocking statement that aims its satiric barbs at African American complacency, southern "justice," southern white gentlemen, and white prostitutes:

> Every Negro paper in this country ought to immediately publish the official records of the Scottsboro cases so that both whites and blacks might see at a glance to what absurd farces an Alabama court can descend (Or should I say an American court?). . . . The 9 boys in Kilby Prison are Americans. Twelve million Negroes are Americans, too. . . . The judge and the jury at Scottsboro, and the governor of Alabama, are Americans. Therefore, for the sake of American justice (if there is any), and for the honor of Southern gentlemen (if there ever were any), let the South rise up in press and pulpit, home and school, Senate Chambers and Rotary Clubs, and petition the freedom of the dumb young blacks—so indiscreet as to travel unwittingly, on the same freight train with two white prostitutes. . . . And, incidentally, let the mill owners of Huntsville begin to pay their women decent wages so they won't need to be prostitutes.[19]

While the first part of Hughes's argument unfolds syllogistically, he seems to be much more interested in taunting his targets than in finding common ground with them. Thus:

If these twelve million Negro Americans don't raise such a howl that the doors of Kilby Prison shake until the 9 youngsters come out (and I don't mean a polite howl, either), then let Dixie justice (blind and syphilitic as it may be) take its course, and let Alabama's Southern gentlemen amuse themselves burning 9 young black boys till they're dead in the State's electric chair. And let the mill-owners of Huntsville continue to pay women workers too little for them to afford the price of a train ticket to Chattanooga. . . . Dear Lord, I never knew until now that white ladies (the same color as Southern gentlemen) traveled in freight trains . . . Did you, world? . . . And who ever heard of raping a prostitute?[20]

In its truculent, uncompromising attitude; its determination to arouse the black community from its complacency and inscribe it with a central role in the struggle against racial injustice; its gestures towards a class analysis of southern customs and attitudes, including an attempt to skewer the "false consciousness" of southern "gentlemen"; and even in its feeble attempts to articulate social distinctions between southern white women "workers" and prostitutes (attitudes shared by many supporters of the Scottsboro Boys), "Southern Gentlemen, White Prostitutes" appropriated many of the elements of the Scottsboro Narrative as it had been constructed by the International Labor Defense and the Communist Party–USA.

Hughes's poem "Christ in Alabama" was equally graphic and provocative:

Christ is a nigger.

Beaten and black:
Oh, bare your back!

Mary is His mother
Mammy of the South,
Silence your mouth.

God is His father:
White master above
Grant Him your love.

Most holy bastard
Of the bleeding mouth,
   Nigger Christ
   On the Cross
   Of the South.[21]

The image of the suffering black male body cast in terms of the Crucifixion is at least as old as Frederick Douglass's account of his savage beating by the slave-master Covey in his 1845 *Narrative*. In "Christ in Alabama" Hughes effectively redeployed this image through bold, assertive, short lines that are unsparing in their indictment of southern racism.

Buttitta and Abernathy timed the release of *Contempo* to coincide exactly with Hughes's appearance in Chapel Hill, and his words immediately triggered off an uproar. As Hughes later recalled in his autobiography, *I Wonder as I Wander*:

> The theatre of the University at Chapel Hill was packed to the doors the night I read my poems there, and special police were on guard to prevent trouble since considerable pressure had been put upon the University to cancel my lecture. Courageously, the University refused to do so. But a leading politician of the town attempted to get police protection for the program withdrawn, stating that I should be run out of town before I had a chance to speak. "It's bad enough to call Christ a bastard," he cried, "but to call him a Nigger—that's too much!"[22]

Buttitta and Abernathy subsequently compounded their assault upon the sensibilities of Chapel Hill and its surrounding communities by inviting Hughes—and several white students who were willing to challenge the racial status quo—to join them for dinner at a public restaurant in Chapel Hill, an event Buttitta gleefully recalled in an article he wrote for Nancy Cunard's *Negro* anthology:

> On that particular occasion we had Langston Hughes for dinner at the snappiest cafeteria in town . . . everyone looked and looked for surprise, and what was the idea of it all . . . the Negro waiters and waitresses were particularly pleased in seeing the event. Much comment arose. We were going to be boycotted for carrying on like this; some people even threatened to make the manager of the cafeteria explain . . . of course, he didn't know . . . then we took Langston to a white drug store for a drink . . . the cheap, southern soda jerker took Hughes for a Mexican or something and let it go at that, but since he found out that Hughes was a "nigger" and he had given him service the way he would have a white man, he got angry and attempted to catch us in a place or two and sock us in the jaw . . . told his friends about it and we were looked at with cheap vengeance for some days. That is not all. We had Langston in our bookshop for a day or so . . . ladies of the courtly order of the University came around to talk and argue with Hughes on the color line, race question and other such stuff . . . a certain wife of a certain nephew of good old Woodrow Wilson—she can't have kids by the way—thought Hughes was darling, cute, and wanted to have him out to dinner in her Dreamship

. . . her home in the Buttons, a section in Chapel Hill, . . . think she for-
got it or changed her mind. . . .[23]

"Scottsboro" and "Christ in Alabama" were subsequently incorporated into
Scottsboro Limited: Four Poems and a Play in Verse. For this volume, Hughes
also dusted off one of his very early poems, "Justice," which was first published in the New York *Amsterdam News* in April 1923:[24]

> That Justice is a blind goddess
> Is a thing to which we black are wise
> Her bandage hides two festering sores
> That once perhaps were eyes.

And he added another aphoristic rhymed couplet that he had originally sent to
his friend Carl Van Vechten in a note from Alabama in early January 1932:

> Scottsboro's just a little place:
> No shame is writ across its face—
> Its court, too weak to stand against a mob,
> Its people's heart too small to hold a sob.[25]

The heart of this volume, however, is *Scottsboro Limited*, a "mass recita-
tion" or "mass chant" inspired by the left-wing German-speaking Proletbuehne, a New York theater group which was credited with introducing a
new type of worker theater with its 1931 production of *Scottsboro*: "the mass
chant, consisting of a simple factual story, or a poem, which builds to a direct
agitational appeal. Recited with tonal effects by a group of workers standing
stock-still—or making simple mass movements—these chants often managed
to lift you off your feet."[26]

As reported in the *Daily Worker*, the chant was first publicly performed in
New York City in an event featuring August Yokinen, a Communist Finnish
immigrant who had been the subject of an elaborate, highly orchestrated, and
well-publicized trial at the Harlem Casino on Sunday, March 1, 1931. Accused of "white chauvinism" after his hostile response to black Communists
seeking admission to a social event at the Finnish Workers Club in Harlem,
Yokinen was tried before an audience of nearly two thousand spectators, found
guilty of "unconscionable chauvinism," and expelled from the Communist
Party. The Party extended him the possibility of rehabilitation, however, if "he
fought racism in the Finnish Workers' Club, joined the League of Struggle for
Negro Rights, sold the *Liberator* on Harlem street corners, and led a demonstration against a local Jim Crow restaurant"—even though, ironically, there
was probably a direct connection between his public trial and the subsequent
decision of the United States government to expel him as an undesirable
alien.[27] His appearance at the *Scottsboro* performance was apparently another
aspect of his penance.

A play based on the Scottsboro frame-up of the nine innocent Negro boys, short speeches by Stephen Graham and August Yokinen, two workers facing deportation to fascist countries, and by William Weinstone will be among the features of the concert and ball tonight in the Manhattan Lyceum of the Committee for the Protection of the Foreign Born.

The prominent Ukrainian basso, Dijo, will sing. The pianist, D. Kotkin, has also been secured to take part in the program with some of his splendid musical selections.

Besides these prominent artists, will be another prominent Ukrainian singer, M. Dmitryshin and the ProletBuhne German workers theatrical group who will give a dramatized story based on the Scottsboro Frame-Up.[28]

The Proletbuehne's production of *Scottsboro* was arranged for six players, who "run from the back of the audience shouting 'Attention, Attention' until they are in their positions."[29] With a minimalist set and stage directions, the production unfolds in a straightforward narrative:

> **ALL:** Attention!
>
> 1: Hear the story of the nine Negro boys in Scottsboro, Alabama
>
> **ALL:** In Scottsboro, in Scottsboro
> Murder is going on.
> In Scottsboro, in Scottsboro,
> Death is threatening. . . .
>
> 2: Everybody is here
>
> 1: on the streets of Scottsboro
>
> 2: The children with flags,
>
> 1: and the men with guns.
>
> 3: And the badges of the sheriffs gleam in the sun
>
> 4: And the brass band plays: Star Spangled Banner!
>
> 2: And everyone is shouting
>
> 3: And everyone is screaming:
>
> **ALL:** Lynch! Lynch! Lynch!. . . .[30]

After swiftly recounting the bare facts of the case, the play offers an explanation for why Scottsboro has occurred:

> 1: Because 12 million Negroes must be kept in slavery

3: By fear, by lack of rights, by violence.

2: For the landlords, bosses and bankers

1: squeeze profits from these 12 million Negroes

2: by long hours, speed up, low wages.

4: If twelve million Negroes are working long hours,

1: also white workers must work long hours.

5: If twelve million Negroes are speeded up,

1: also white workers are speeded up.

6: If twelve million Negroes work for low wages,

1: also white workers must work for low wages.

3: Negro workers

1: White workers

4: The same exploitation

5: The same fate

6: The same class.

3: Negro workers

1: White workers

2: The same power

3: The same might

1: The same certainty of victory

4,5,6: If you unite

1,2,3: If you organize

ALL: IF YOU FIGHT. . . . [31]

"Scottsboro" pitches directly and inevitably forward to its final appeal:

ALL: STOP! STOP! STOP!

1: Fight for the liberation of the nine Negro boys

2: is a part of the class struggle,

5: led by the workers of all countries

1: against the bosses of all countries

ALL: WORKERS, UNITE AND FIGHT![32]

Against the backdrop of the Proletbuehne's *Scottsboro*, Hughes's *Scottsboro Limited* can be seen as a subtle and innovative appropriation of this dramatic form. Like *Scottsboro*, *Scottsboro Limited* uses a bare stage for its setting, but it revolves around a much larger cast: eight black boys, two white women, a white man, eight white workers, and voices in the audience. From the very beginning, the eight black boys are central to the action of the play: they are not spoken about, they speak. As they approach the stage, chained to each other, their presence is challenged by a white man in the audience:

> **MAN:** (*As the Boys mount the stage, the Man rushes up to them threateningly.*) What the hell are you doing in here, I said?
>
> **1ST BOY** (*turning simply*):
> We come in our chains
> To show our pain.
>
> **MAN** (*sneeringly*): Your pain?! Stop talking poetry and talk sense.
>
> **8TH BOY** (*as they line up on the stage*):
> All right, we will—
> That sense of injustice
> That death can't kill.
>
> **MAN:** Injustice? What d'yuh mean? Talking about injustice, you coon?
>
> **2ND BOY** (*Pointing to his comrades*):
> Look at us then:
> Poor, black, and ignorant,
> Can't read or write—
> But we come here tonight.[33]

As James Smethurst has astutely observed, the different levels of language usage among the characters in the play are important clues to its development and political meaning.[34] The White Man and the voices of the southern white mob utter their crude, prosaic sentiments in southern vernacular speech, while the Boys speak in simple but elegant poetry. The play unfolds through flashbacks, as the Boys recount their story of their fateful ride on a freight car;

their arrest; the discovery of two white women on the same freight car, and the false accusation of rape; their trial, conviction, and impending execution. In the early stages of their story, the voices of the Boys are represented by rural, African American vernacular speech, even as they express pre–"class-conscious" insights into their condition as young, black southern men:

> 2ND BOY: All them little town jobs we used to do,
> Looks like white folks is doin' 'em now, too.

> 1ST BOY: Just goes to prove there ain't no pure nigger work.

> 6TH BOY (*in wonder*):
> Look a-yonder you'all, at dem fields
> Burstin' wid de crops they yields.
> Who gets it all?

> 3RD BOY: White folks.

> 8TH BOY: You means de rich white folks.

> 2ND BOY: Yes, 'cause de rich ones owns de land.
> And they don't care nothin' 'bout de po' white man.

> 3RD BOY: You's right. Crackers is just like me—
> Po' whites and niggers, ain't neither one free.[35]

In the aftermath of their trial and conviction—complete with cries by an angry mob for their deaths—the Boys sit in the death house contemplating their fate:

> 2ND BOY: But how you gonna git out o' here
> With these iron bars and this stone wall
> And the guards outside and the
> Guns and all?

> 8TH BOY: There ain't no way for a nigger to break free.
> They got us beaten, and that's how we gonna be
> Unless we learn to understand—
> We gotta fight our way out like a man.

> 5TH BOY: Not out of here.

> 4TH BOY: No, not out o' here.
> Unless the ones on the outside
> Fight for us, too,
> We'll die—and then we'll be through.

This is the critical moment in the play when the Red voices in the audience begin to murmur:

We'll fight for you, boys. We'll fight for you.
The Reds will fight for you.[36]

Beyond this point there is a direct relationship between the rapid development of the Boys' political awareness and their mode of speech—culminating
with the ideological and physical eruption of the 8th Boy:

Burn *me* in the chair?
NO!
(He breaks his bonds and rises,
tall and strong.)
NO! For me not so!
Let the meek and humble turn the other cheek—
I am not humble!
I am not meek!
From the mouth of the death house
Hear me speak![37]

The 8th Boy triumphantly announces the arrival of the "New Red Negro,"
whose voice is virtually indistinguishable from those of the Red voices who
support them. Together they smash the electric chair, tear down the bars that
confine the Boys (and separate them from their working-class allies), and lead
the audience in a rousing chant:

BOYS AND REDS: All hands together will furnish the might.

AUDIENCE: All hands together will furnish the might.

RED VOICES: Rise from the dead, workers, and fight!

BOYS:
Together, black and white,
Up from the darkness into the light.

ALL: Rise, workers, and fight!

AUDIENCE: Fight! Fight! Fight! Fight!

(*The curtain is a great red flag rising to the strains of the Inter-
nationale.*)[38]

Hughes wrote and published *Scottsboro Limited* before he actually met the
Scottsboro Boys in late January 1932, during the leg of his southern reading
tour when he appeared at Tuskegee Institute. Visiting them one Sunday in
Kilby Prison, on the outskirts of Montgomery, Hughes must have been struck
by the sharp discrepancy between the heroic defiance he conferred upon
them in *Scottsboro Limited* and their actual grim and dismal circumstances.
Of his visit Hughes later wrote:

Their chaplain, a small-town Negro minister, said it might cheer the boys up if I would read them some of my poems. So at Kilby Prison I went down the long corridor to the death house to read poetry to the Scottsboro boys. In their grilled cells in that square room with a steel door to the electric chair at one end, in their gray prison uniforms, the eight black boys sat or lay listlessly in their bunks and paid little attention to me or to the minister as we stood in the corridor, separated from them by bars. Most of them did not even greet us. Only one boy came up to the bars and shook hands with me.[39]

Hughes must have vicariously experienced the spirit-crushing weight of their confinement, too. An essay published in *Opportunity* five months after his visit vividly captured his own sudden rush of terror inside Kilby Prison: "For a moment the fear came: even for me, a Sunday morning visitor, the doors might never open again. WHITE guards held the keys. . . . and I'm only a nigger."[40] Hughes opted for an upbeat performance of his poetry, his militant rhetoric and politics apparently collapsing under the weight of these extreme conditions:

I selected mostly my humorous poems to read to them, and I did not mention the South or their problems, because I did not know what to say that might be helpful. So I said nothing of any seriousness, except my hope that their appeals would end well and they would soon be free. The youngest boy, Andy Wright [whom Hughes apparently confused with Eugene Williams], smiled. The others hardly moved their heads. Then the minister prayed, but none of the boys kneeled or even changed positions for his prayer. No heads bowed.[41]

Scottsboro would continue to haunt Hughes throughout the decade; he revisited the subject in the early summer of 1938 with "August 19th . . . A Poem for Clarence Norris." Norris was one of the Scottsboro defendants first sentenced to death in 1933, but granted a reprieve by the Supreme Court of the United States in 1935. The famous ruling *Norris v. Alabama* ruled that qualified blacks had been systematically excluded from jury duty in the first Alabama trial. In July 1937 Norris was tried again, found guilty again, and again sentenced to death. By this date, Norris was the only Scottsboro defendant still facing the death penalty; ironically, during the same month the Alabama authorities dropped all charges against Eugene Williams, Olen Montgomery, Willie Roberson, and Roy Wright. When Hughes originally published "August 19th . . ." in the *Daily Worker* in 1938, he appended the following note: "Read this poem aloud, and think of young Clarence Norris pacing his lonely cell in the death house of Alabama, doomed to die on August 18. When used for public performances, on the last two verses punctuate the poem with a single drumbeat after each line. AUGUST 18TH IS THE DATE. During the final stanza, let the beat go faster, and faster following the line, until at the end the drum goes on alone, unceasing, like the beating of a heart":[42]

> What flag will fly for me
> When I die?
> What flag of red and white and blue,
> Half-mast against the sky?
> I'm not the President,
> Nor the honorable So-and-So.
> But only one of the
> Scottsboro Boys
> Doomed "by law" to go.
> **August 19th is the date.**[43]

As in *Scottsboro Limited*, Hughes writes from the perspective of the Boys themselves—in this case the figure of Clarence Norris, conferring upon him an inner life sometimes overlooked by the fervent supporters of the Scottsboro defendants. In its irregular lines and rhyme scheme, and its ordinary but elegant diction, "August 19th . . ." effectively captures the pace of a rapid heartbeat, particularly in the antiphonal lines that end the poem:

> AUGUST 19th IS THE DATE.
> *Can you make death wait?*
> AUGUST 19th IS THE DATE.
> *Will you let me die?*
> AUGUST 19th IS THE DATE.
> *Can we make death wait?*
> AUGUST 19th IS THE DATE.
> *Will you let me die?*
> AUGUST 19th IS THE DATE.
> AUGUST 19th IS THE DATE.
> AUGUST 19th . . . AUGUST 19th . . .
> AUGUST 19th . . . AUGUST 19th . . .
> AUGUST 19th . . .[44]

Many of the tropes of Hughes's poems were echoed in the work of Herman J. D. Carter, a creative writing student of James Weldon Johnson at Fisk University in Nashville, Tennessee, whose poems circulated in the African American press. His 1933 collection *The Scottsboro Blues* took him on a reading-speaking tour to Chicago and other cities during the mid-1930s and was "dedicated to the nine Scottsboro boys with the hope that the sale of it might assure them a small part of the financial support necessary to rid them of the stain which has been spilled on their souls."[45]

The collection emerged, Carter declaims, "out of that deep fuliginous sea of desperation made so by Fate's uncanny, oppressive envoy, Prejudice, who landed an everlasting blow on the souls of twelve-million blacks, with such terrific force that the echo was heard around the world."[46]

The title poem makes effective onomatopoetic use of the image of the slave-master's whip, reinforcing the depiction of black degradation and victimization:

> *Swish! slash!*
> *Swish! slash! pop!*
> "Oh Lawdy!. . . . *Swish!*
> Ma-a-a Lawdy. . . . *slap!*
> Don' let cap'en
> Hit me 'g'in. . . . *swish!*[47]

Consistent with the Christian outlook that shapes the poem, the speaker cries out for mercy, not for himself but for his captor:

> Lawdy, hab mucy
> On dey souls. . . . *swish!*
> Hab mucy on
> Deese,. . . . *pop!*
> Ma white 'bacco chewin'
> Massers. . . . *slap!*
> Hab mucy Lawd . . . *swish!*[48]

In "A Voice from Scottsboro," Carter also extends Christian forgiveness to Ruby Bates, who recanted her destructive testimony during the second Scottsboro trials in Decatur, Alabama in 1933—even as the speaker of the poem maintains a censorious attitude towards her sexual promiscuity:

> Ruby Bates,
> Hush yo' cryin'
> I kno's I's gotta die
> Fo' whut I didn't do;
> But don' yu cry no mo';
> 'Cause my blood ain' 'quired at yo' han's.
> Yu tol' de truf;—
> Wise men kno' de truf whey dey hure 't;
> Fools don't.
> Yu ain' no murderer,
> Fools, cornfield, red-neck fools
> Iz murders, not yu;
> So,
> Hush yo' cryin' Ruby.

I kno's hi comes yi grieves so;
'Cause furst yu lied on me.
Das alright,
Me 'n' God fo'gives yu fo' dat;
'Cause a harlot made yu do't.
Yu's ah harlot too Ruby.
But yu's one whuts got pride.
Hush ni Ruby!
Dry up!
Yu done yo' part
An' failed;
Ni God iz gonna do His.[49]

This theme of Christian redemption achieves its apotheosis in "The Blacker Christ," which explicitly locates Scottsboro defendant Haywood Patterson at the end of a range of saviors, a necessary sacrifice to appease southern bloodlust:

"Haywood, Haywood!"
Called a voice,
A voice from the sky,
"I gave my son to man;
He died that you might
Live; but
You were again enslaved.
I sent Abe Lincoln
To set you free; again
You were enslaved.
My son died for you;
Lincoln died for you, —
Still my people are
Slaves.
Haywood, die!!!
Your people need
Your blood,
Then
Southern whites
Won't be so quick
To
Put a black in chains;
White women,
Before they scream
Will use their feeble brains.
Haywood, my son,

Die!!!
Your body will die; but
Not your soul; 'cause
You're my son,
The Blacker Christic.[50]

Samuel Leibowitz, the lawyer hired to defend the Scottsboro Boys by the ILD, makes a cameo appearance in this volume ("Leibowitz, New Christ / Takes my stand") as does the southern attorney George Chamlee ("Link me not with / *Injustice*").

In spite of the mood of Christian fatalism that pervades this sequence of poems, they end on a surprisingly upbeat note, when a southern policeman disrupts the threats of two white women to accuse black men driving on a rainy country road with rape unless they give them a ride. He brusquely intervenes, sending the women on their way:

"Git on down de road,
Ah done heard it all."

T'was an honest white cop
Hid in the road's mouth.
I pray my God for more honest white cops,
To hide in the bushes of the South.[51]

Although she was twelve years younger than Langston Hughes, Muriel Rukeyser, too, was swept up in the public spectacle of the Scottsboro case. Born in 1913 to an upwardly mobile Jewish family, Rukeyser grew up on New York's Riverside Drive and was educated at the Fieldston Schools and Vassar College. But an early poem, "Poem out of Childhood," evokes a world permanently marked by the devastation of World War I, Sacco and Vanzetti, the Depression, and other unnamed "horrors." From her childhood, it seems, Rukeyser was drawn to the world of public events, compelled by an abiding vision of social justice. Her father's bankruptcy forced her to leave Vassar after her sophomore year (and, according to critic Louise Kertesz, "he later disinherited her for her political views and 'disobedience'"[52]). By then, however, she had already clearly chosen the artistic and political path that marked a sharp break with her social background. In 1933, Rukeyser drove to Decatur, Alabama to report on the second Scottsboro trials for the left-wing magazine *Student Review*, "but she was arrested by the police when they found her and two friends talking to black reporters and discovered in her valise thirty posters advertising a black student conference at Columbia."[53] Rukeyser contracted typhoid fever while she was being held in an Alabama jail, but she regarded this affliction as minor in the face of the magnitude of the injustice she witnessed there. Out of this experience Rukeyser crafted the powerful poem about the Scottsboro Boys that appeared in her first volume of poetry, *Theory of Flight*, the collection that won her the Yale Younger Poets Prize in 1935.

Subdivided into three sections, "The Lynchings of Jesus" appropriates patterns of crucifixion imagery already familiar to readers of poems like Langston Hughes's "Christ in Alabama." And, like Hughes in his poem "Scottsboro," Rukeyser sees revolutionary potential in any figure who strives to challenge the limitations of his environment—beginning with the martyred Jesus Christ. Thus, in the first section of the poem, "Passage to Godhead," Jesus is depicted as an emblematic revolutionary, whose struggles assure his immortality:

> Passage to godhead, fitfully glared upon
> by bloody shinings over Calvary
> this latest effort to revolution stabbed
> against a bitter crucificial tree,
> mild thighs split by the spearwound, opening
> in first gestation of immortality.[54]

Jesus stands at the head of a long line of revolutionary victim-heroes:

> Many murdered in war, crucified, starved,
> loving their lives they are massacred and burned,
> hating their lives as they have found them, but
> killed while they look to enjoy what they have earned.[55]

Bruno, Copernicus, Shelley, and Karl Marx—"makers of victories for us"—provide us with both a revolutionary legacy and the will and the courage to discover heroes in our everyday lives:

> moved to find
> what numbers of lynched Jesuses have not been deified.[56]

The second section of the poem, "The Committee-Room," shifts its attention to the powerful figures who legislate America's destiny but who have no real contact with the people outside of the space they occupy:

> Let us be introduced to our superiors, the voting men.
> They are tired ;   they are hungry ;   from deciding all day
> around the committee-table.
>   Is it foggy outside?   It must be very foggy
>   The room is white with it.[57]

Against this backdrop the poet draws a sharp contrast between the "flab faces" that populate the Committee-Room and those whose sense of language and experience derive from more passionate and committed traditions, among them her Jewish ancestors:

> Those people engendered my blood swarming
> over the altar to clasp the scrolls and Menorah
> the black lips, bruised cheeks, eye-reproaches :
> as the floor burns,° singing Shema[58]

The clarity—indeed, the purity—of the traditions represented by "the scrolls and Menorah" stands in sharp contrast to the corruption of language represented by contemporary journalistic practices:

> Our little writers go about, hurrying the towns along,
> running from mine to luncheon, they can't afford
> to let one note escape their holy jottings:
> today the mother died, festering : he shot himself : the bullet entered
> the roof of the mouth, piercing the brain-pan[59]

This corruption of language, in turn, is the basis of a mangled and distorted historical consciousness, as we witness in the following exchange between mother and child:

> What did you do in school today, my darling?
>   Tamburlaine rode over Genghis had a sword
>   holding riot over Henry V Emperor of and
>   the city of Elizabeth the tall sails
>   crowding England into the world and Charles
>   his head falling many times onto a dais
>   how they have been monarchs and
>   Calvin Coolidge who wouldn't say
>   however, America[60]

From the image of "Silent Cal" Coolidge, who apparently will not—or cannot—reveal anything about the deepest truths of American life, the poem cuts back to the Committee-Room, where its members not only cast votes on arbitrary and trivial matters, but also condemn the rebels, the visionaries, and the powerless:

> We are powerful now : we vote
>   death to Sacco a man's name
>   and Vanzetti a blood-brother; death
>   to Tom Mooney, or a wall, no matter;
>   poverty to Piers Plowman, shrieking
>   anger
>   to Shelley, a cough and Fanny to
>   Keats;
>   thus to Blake in a garden; thus to
>   Whitman;

thus to D.H. Lawrence.
                    And to all you women,
dead and unspoken-for, what
        sentences,
to you dead children, little in the
        ground
   : all you sweet generous rebels what
        sentences[61]

In anticipation of the scene we will encounter in the third part of the poem, the poet directs our attention to the grisly accounts of an 1897 lynching in Texas:

This is the case of one Hilliard, a native of Texas,
in the year of our Lord 1897, a freeman.
Report. . .Hilliard's power of endurance seems to be
the most wonderful thing on record.   His lower limbs
burned off a while before he became unconscious:
and his body looked to be burned to the hollow.
Was it decreed (oh coyly coyly) by an avenging God
as well as an avenging people that he suffer so?
                                We have
16 large views under magnifying glass.
8 views of the trial and the burning.
For place of exhibit watch the street bills.
            Don't fail to see this.[62]

The final sequence of "The Committee-Room" presents two clashing images. A Committee-member returns home in the evening, weary after a long day of voting upon the lives of other people, and receives solace from a supportive and attentive spouse. Counterposed to this scene, however, the concluding stanza conjures up the growing murmurs of a gathering crowd and a rising storm:

all night they carried leaves
bore songs and garlands up the gradual hill
the noise of singing kept the child awake
    but they were dead
all Shakespeare's heroes   the saints   the
        Jews   the rebels
    but the noise stirred their graves' grass
    and the feet all falling in those places

> going up the hill with sheaves and tools
> and all the weapons of ascent together.[63]

Outside of the United States the Scottsboro case commanded the attention of literally millions of people from across the globe, who sent telegrams, letters, and petitions. Acts of poetic witnessing, encouraged by countless literary competitions in the European press inspired by the 1932 tour of Scottsboro mother Mrs. Ada Wright and the ILD's Louis Engdahl,[64] occurred across the continent—such as "Are We to Burn," a poem that appeared in the *Trybuna*, the Berlin Red Aid magazine:

> How the time flies! How the time
> flies!
> In front of our cells, in front of
> our eyes
> Is the electric chair.
> The jailers torture us,
> "Nigger,
> look,
> That's the way you, and you, will
> Cook
> In the electric chair. . . ."
> All know it is not for "alleged at-
> tack"
> But for being workers and being
> black
> That we go to the Chair. . . .
> But all the workers, black and
> white,
> Are learning to organize and fight
> You and your Chair.
> Are we to burn? Are we to die?
> Workers, speak! Make your reply
> To them, and their chair—
> They Shall Not Die![65]

Writer/activists such as these engaged in mass demonstrations far removed from the jaded world of English Communists retrospectively portrayed by Nancy Mitford in her 1945 novel *The Pursuit of Love*:

"The worst of being a Communist is that the parties you may go to are— well—awfully funny and touching, but not very gay, and they're always in such gloomy places. Next week, for instance, we've got three, some Czechs at the Sacco and Vanzetti Memorial Hall at Godlers Green, Ethiopians at the Paddington Baths, and the Scotsboro (sic) boys at some boring old rooms or other. You know."

"The Scotsboro [sic] boys," I said. "Are they really still going? They must be getting on."

"Yes, and they've gone downhill socially," said Linda, with a giggle. "I remember a perfectly divine party Brian gave for them—it was the first party Merlin ever took me to so I remember it well, oh dear, it was fun. But next Thursday won't be the least like that. . . .I am always saying to Christian how much I wish his buddies would either brighten up their parties a bit or else stop giving them, because I don't see the point of sad parties, do you? And left-wing people are always sad because they mind dreadfully about their causes, and the causes are always going so badly. You see, I bet the Scotsboro boys will be electrocuted in the end, if they don't die of old age first, that is. One does feel so much on their side, but it's no good. . . ."[66]

In the English context, one of Nancy Mitford's satiric targets likely would have been Nancy Cunard, the flamboyant and uncompromising social rebel who plunged wholeheartedly in the defense of the Scottsboro Boys. Born in England in 1896 to American-born Maud Burke and Sir Bache Cunard, a grandson of Samuel Cunard, the founder of the famous steamship line, Cunard accompanied her mother to London after her parents divorced. Cunard soon found her place among the literary and political bohemians of post–World War I Paris, publishing three volumes of poetry in quick succession. In 1926, on an extended trip to Venice, Cunard met a group of African American musicians, the first African Americans she had ever known—among them a pianist named Henry Crowder. Her meeting with Crowder marked the beginning of Cunard's extensive involvement in issues of race, racism, and African American life. In 1928 Crowder moved with her to her new country home in Normandy, where he assisted her in the establishment of her own printing operation, Hours Press.

By the fall of 1930, Cunard had decided to organize a comprehensive anthology of black history, politics, and culture. She and Crowder embarked for London to collect material, and it was just a matter of time before her mother, Maud, learned that she was living with a black man. Maud was appalled and confronted her daughter. Cunard was adamant in her refusal to renounce Crowder. The meeting ended in a shambles, with Cunard vowing never to speak to her mother again. The English popular press began printing sensational stories of the visit of Cunard and Crowder to England, provoking a spate of insulting and threatening letters to the besieged couple. Maud hired detectives to follow her daughter, and she put pressure on the man who had provided the couple with lodgings. Harassed and exhausted, Cunard and Crowder returned to the Continent.

In the aftermath of this experience Cunard wrote and had privately published an eleven-page pamphlet, *Black Man and White Ladyship: An Anniversary*, a savage broadside that mercilessly pilloried her mother and her social circle, and vented Cunard's hatred of racism on a global scale. This episode represented the final rupture of her relationship with her mother and stiffened her resolve to produce her anthology—a resolve that was heightened by her deepening involvement with the Scottsboro case.[67]

While working on the anthology, Cunard made two visits to the United States, one in 1931, the other in 1932. On the second visit, when she was accosted by American reporters pursuing rumors that she was romantically linked to Paul Robeson, she delivered the following speech:

> And now after your interest in my private affairs . . . I want something in return. I want money for the defense of the 9 Negro framed-up boys held under death sentence since a year, the Scottsboro, Alabama case. I ask for money to be sent immediately to me for the Scottsboro defense. *Why are you Americans so uneasy of the Negro race?*. . . . Who'll write me the best answer to this? I'll print it in my book on *Colour* [*her original title for* Negro].

In this manner Cunard publicly identified herself to an American audience as a passionate defender of the Scottsboro Boys.

Back in England, Cunard served as honorary treasurer of the British Scottsboro Defence Committee, which organized demonstrations, benefit evenings, films, and petitions. She contributed personal funds, sometimes sending small amounts to the families of the defendants, bullied and cajoled her friends for cash and signatures, and marched in demonstrations. After one such event in London she elaborated on her views to a reporter for the *Daily Express*, who was as much taken with her physical appearance as he was with her unconventional political views:

> Miss Nancy Cunard's exotically massive slave bangles jangled on her thin white arms; her black-circled eyes dashed indignantly; and above the cosmopolitan babel of the noisy West-end café in which we sat last night she seemed to shout, yet only whispered:
> "*I hate so many white people.*"[68]

Asked why she was so deeply invested in the "cause of negroes," Cunard responded:

> because they are the most oppressed race of civilized people in the world to-day. . . . I like them because I seem to understand them. My mother is an American, but she does not understand them. . . .
> This Scottsboro case is one of the most appalling miscarriages of justice that a so-called civilized country has ever perpetrated. . . .

How can you talk about civilization…when you sit back and scoff at such injustice done to men born into this world just because they are darker skinned than we are.[69]

Indeed, her anthology *Negro* needs to be understood as a monument of her emotional and political commitment to the Scottsboro case.

Published in February 1934, *Negro* was a massive anthology—855 pages long, two inches thick, almost eight pounds in weight—that included about two hundred and fifty contributions by some one hundred and fifty authors, two-thirds of them people of color from across the world. Sprawling, eclectic, and idiosyncratic, over half of *Negro* examined aspects of black life in the Americas; another three hundred pages dealt with Africa; sixty were devoted to Europe.

For all of its sprawl, however, Cunard left no doubt about the political sympathies that guided her project: "It was necessary to make this book…for the recording of the struggles and achievements, the persecutions and revolts against them of the Negro peoples."[70] At the heart of the anti-racist and anti-imperialist politics of *Negro* stood her essay "Scottsboro—and other Scottsboros," the most sustained piece of political journalism Cunard produced during her lifetime and the most detailed account of the case—heavily embroidered with passionate partisan rhetoric, to be sure—up until September 1933.

Drawing heavily upon published accounts of the case in England's *Daily Worker*, Cunard hewed closely to the rhetoric of the International Labor Defense in her editorial interventions in the article:

> There are certain sections of the Negro bourgeoisie which hold that justice will come to them from some eventual liberality in the white man. But the more vital of the Negro race have realized that it is Communism alone which throws down the barriers of race as finally as it wipes out class distinctions. The Communist world-order is the solution of the race problem for the Negro.[71]

In the ongoing war of words and behind-the-scenes struggle for control of the Scottsboro case between the International Labor Defense, the Communist Party, and the NAACP, Cunard declared herself squarely on the side of the International Labor Defense—but her point of view did not prevent her from publishing articles in *Negro* by W.E.B. DuBois, William Pickens, and other NAACP supporters whose views stood in uneasy relationship with Cunard's relentless critique:

> I have myself had to listen to a partisan of this organization actually assert that had the Scottsboro boys pleaded guilty they might have gotten life-terms, "and then they might have been set free, oh, some time later." It is this odious and vicious treachery of the black bourgeoisie in America (as

elsewhere), hand in glove with the lynchers and oppressors of black people, that makes the fight for their liberation doubly long and bitter. It has been shown how world-protest and exposure of the Scottsboro trial scandal has won appeals and re-trials—partial victories only so far, but of vital importance not only to the innocent boys themselves but to the future of the whole Negro race in the U.S. The N.A.A.C.P. is against mass pressure and protest; it does not believe in the solidarity of black and white workers and other militant sympathizers.[72]

Buttressed by voluminous evidence in the form of ILD press releases, personal correspondence, secondhand accounts, and stories from the U.S. press, Cunard hammered away at her point—painting in vivid prose and dramatized descriptions the plight of the Scottsboro Boys:

And the boys in the rotten jail you could break out of with a spoon. . . . A special correspondent went to see them. Thirteen iron steps and the shadow of the hangman's noose. So, for 2 years they'd been face to face with the electric chair across the passage in front of their cells and now they had the gallows to stare at in the electric light. On the last step was the trap-door to eternity—a painting of Christ at Gethsemane, done by some dead convict above it. The correspondent looked at the gallows and the Christ. Then he saw the eyes—the Scottsboro boys were looking at him. It was full of shadows in that place—jailors and guards moving about, the shadow of the actual noose, shadows of themselves shuffling round, straining ears for the rumble of mobs. They didn't sleep; they were wondering all the time if they would die in Decatur, lynched, or be convicted again. Did they dare even to think of the possibility of acquittal? Vermin was on them, the cockroaches crept over the filthy cell floor; the bed bugs lived in their clothes, they couldn't get them out; jail gives no change of clothes. "What'll you do when you are freed?" asked the journalist. Olin Montgomery seized one of the greasy bars and leaned against it, peering intently with his one nearsighted eye, "Go home," he said simply. "Oh boss, I wants to go home. I been here 2 years and 9 days, 2 years and 9 days."[73]

Although Cunard did not make the "Scottsboro Pilgrimage" favored by some artists and intellectuals during the 1930s, never journeying South of the Mason-Dixon line during her two sojourns in the United States, her work on behalf of the Scottsboro defendants nevertheless constituted an important act of witnessing that reverberated with those who heard her speeches and read her words—in the United States and abroad. Her efforts certainly resonated with her longtime friend Kay Boyle, an American expatriate writer whose social and literary circles considerably overlapped with those of Nancy Cunard.

Born in St. Paul, Minnesota in 1902, the daughter of Howard Boyle, a lawyer, and Katherine Evans, a literary and social activist, Boyle grew up in a comfortable, upper-middle-class environment, traveling extensively with her family and studying sporadically at institutions ranging from the Cincinnati Conservatory of Music and the Ohio Mechanics Institute to Parson's School of Fine and Applied Arts in New York City and Columbia University. In 1922 she moved in with her sister Joan in New York City, where she worked with Lola Ridge on editing the experimental literary magazine *Broom*—the beginning of her lifelong involvement with the literary avant-garde. She also married French-born engineer Richard Brault, with whom she moved to France in 1923. There she began to work with Ernest Walsh, editor of *This Quarter*. By the mid-1920s, Boyle's marriage to Brault had collapsed and she moved to Grasse, France to live with Walsh. Walsh died in 1927 and Boyle, pregnant with his child, moved to Paris, where she quickly became active in the Paris avant-garde scene, signing the "Revolution of the Word" manifesto issued by Eugene Jolas's magazine *Transition* and publishing frequently in its pages. In 1932 Boyle married Lawrence Vail. During the decade of their marriage, they lived not only in France, but also in Austria and England—from which vantage point Boyle poured out fiction and poetry that was prescient about the rise of Nazism long before it was widely regarded as a problem in the United States and that was adamant in its insistence that the artist must be engaged in the social and political issues of her times.[74]

Published in the June 9, 1937 issue of the *New Republic*, more than a year after the state of Alabama had tried Haywood Patterson for the *fourth* time, convicted him, and sentenced him to seventy-five years, "Communication to Nancy Cunard" was a fitting rejoinder to Cunard's impassioned appeal in "Scottsboro—and Other Scottsboros."

Like Cunard's article, Boyle's poem bears witness to the ordeal of the Scottsboro Boys from afar, at the same time that it, ironically, interrogates the role of the writer and raises fundamental questions about the power of language to ameliorate their condition. Like "Scottsboro—and Other Scottsboros," "Communication to Nancy Cunard" draws heavily upon documentary sources, most notably Carleton Beals's account of the fourth Scottsboro trial that appeared in the *Nation* magazine on February 5 and February 12, 1936, and also upon her own correspondence with Scottsboro defendant Haywood Patterson.[75]

From the beginning of the poem, the speaker is extremely scrupulous about establishing who the poem does *not* speak for:

> These are not words set down for the rejected
> Nor for outcasts cast by the mind's pity
> Beyond the aid of lip or hand or from the speech
> Of fires lighted in the wilderness by lost men
> Reaching in fright and passion to each other.
> This is not for the abandoned to hear. . . .[76]

The "rejected," the "outcasts," and the "abandoned" who are ostensibly excluded from the purview of the speaker include the Scottsboro accusers,

Victoria Price and Ruby Bates, as well as the Scottsboro Boys themselves—all of whom exist in conditions that cannot be mitigated by "the mind's pity" and "beyond the aid of lip or hand." They have been cast outside the realm of ordinary discourse, and speech(es) cannot necessarily alter their circumstances. The speaker also implies that pity for them does not in itself constitute an adequate response, for the abandoned figures of the Scottsboro case inhabit a concrete universe of suffering and deprivation far removed from those who have the freedom to contemplate their fate from the comfort of their bourgeois existences. As Ellen McWhorter has pointed out: "Boyle implicitly critiques armchair theorizing about racial politics, including a critique of poets who do not acknowledge the limitations of working in a medium so fraught with abstract and tangible privilege."[77]

The poem is *for*, the speaker underscores, "the sheriff with the gold lodge pin"—a central figure in Carlton Beals's account of the fourth trial—and "the jury venireman" who discourses at length on the convoluted racist myth, based upon biblical sources, about "how the nigger race begun" with the act of coupling between Cain and a "female baboon / or chimpanzee or somethin' like it" in the land of Nod. It is also for the tobacco-chewing figure popularized by Samuel J. Leibowitz's angry outburst to the press after the 1933 Decatur trials, this time in the form of "the Sunday-school teacher with the tobacco plug / Who addressed the jury, the juice spattering on the wall," who also invoked biblical authority in his address to the jury:

> Now even dogs choose
>   their mates,
> But these nine boys are lower than the birds of the air,
> Lower than the fish in the sea, lower than the beast of the fields.
> There is a law reaching down from the mountaintops to the
>   swamps and caves—
> It is the wisdom of the ages, there to protect the sacred parts of
>   the female Species
> Without them having to buckle around their middles
> Six shooters or some other method of defense.

And the poem is for the Alabama townspeople, "people who go and come, / Open a door and pass through it, walk in the streets / With the shops lit, loitering, lingering, gazing." And for "Deputy Sheriff Sandlin, Deputy Sheriff / Blacock, / With Ozie Powell, handcuffed. Twelve miles out of Cullman / They shot him through the head"—although the speaker neglects to mention that Powell, in spite of his handcuffs, somehow managed to slash Deputy Blacock's throat while he was being transported back to the Birmingham jail in 1936. But it is not for these characters in the sense that it does not attempt to speak to them, or to challenge their intact and provincial views.

As the poem unfolds, it becomes increasingly clear that the speaker is concerned with the complicated relationship between language and power: all of the figures the poem is for are linked by the relatively privileged positions they occupy in a society whose norms are unquestioned and enshrined in a rhetoric buttressed by biblical authority. Victoria Price, Ruby Bates, and the Scottsboro Boys are victimized in part, the speaker makes clear in the second section of the poem, "Testimony," by their problematic relationship to language, dramatically captured by the juxtaposition of the words of Haywood Patterson and Victoria Price. Here Boyle carefully reproduces one of Patterson's letters (Patterson had corresponded with Boyle and her family in the period preceding the poem), complete with all of its spelling and grammatical errors:

> "So here goes an I shell try
> Faitfully an I possibly can
> Reference to myself in particularly
> And concerning the other boys
>   Personal pride
> And life time upto now.
> You must be patiene with me
>   and remember
> Most of my English is not of
>   much interest
> And that I am continually
> Stopping and searching for the
> word."

Patterson proceeds to narrate a story of accident and happenstance, of how the realities of the Depression led him to misfortune "without a moving cause." Next to Patterson's poignant narrative, Victoria Price's *refusal* to testify—"I cain't remember"—stands as a kind of choral refrain. Many readers in the 1930s would have quickly grasped the irony of this juxtaposition, since it was Victoria's convenient but consistent lapses of memory through all of the Scottsboro trials that helped to secure the boys' convictions. Nonetheless, the speaker, in the final analysis, refuses to make sharp distinctions between her and the Scottsboro Boys: in the terms of the poem, Price, Patterson, and, indeed, all of the hoboes riding the rails are victims of the same impersonal economic forces:

> Hear how it goes, the wheels of it traveling fast on the rails The
>   boxcars, the gondolas running drunk through the night. . . .
> Here it passing in no direction, to no destination
> Carrying people caught in the boxcars, trapped on the coupled
>   chert cars. . . .
> Not the old or the young on it, nor people with any difference in
>   their color or shape,

> Not girls or men, Negroes or white, but people with this in
>   common:
> People that no one had use for, had nothing to give, no place to
>   offer
> But the cars of a freight train careening through Paint Rock,
>   through Memphis,
> Through town after town without halting.
> The loose hands hang down, and swing with the swing of the
>   train in the darkness,
> Holding nothing but poverty, syphilis white as a handful of dust,
>   taking nothing as baggage

The next section of the poem, "The Spiritual For Nine Voices," offers a sharp satirical portrait of the solace offered the Scottsboro Boys by otherworldly Christianity, a vision perhaps inspired by Marc Connelly's film adaptation of his Pulitzer Prize–winning play *The Green Pastures*, interestingly enough in 1936. Featuring an all-black cast that included the talents of Rex Ingram, Edna Mae Harris, and Eddie "Rochester" Anderson, *The Green Pastures* reenacted stories of the Old Testament, depicting Heaven as an extended community fish-fry, attended by everyone, including God himself. The rhythms and diction of this section of the poem take on a gospel beat, with undertones of minstrelsy:

> I went last night to a turkey feast (Oh, God, don't fail your chil-
>   dren now!)
> My people were sitting there the way they'll sit in heaven
> With their wings spread out and their hearts all singing
> Their mouths full of food and the table set with glass
> (Oh, God, don't fail your children now!). . . .
>
> There were oysters cooked in amplitude
> There was sauce in every mouth.
> There was ham done slow in spice and clove
> And chicken enough for the young and old. . . .

The parenthetical references to God, however, underscore his absence during crucial moments in the Scottsboro case:

> (Were you looking, Father, when the sheriffs came in?
> Was your face turned towards us when they had their say?). . . .
>
> (Was it you stilled the waters on horse-swapping day
> When the mob came to the jail? Was it you come out in a long
>   tail coat
> Come dancing high with the word in your mouth?)

And the final lines of the poem emerge from the voice of Olen Montgomery:

> Half-blind in oblivion saying: "It sure don't seem to me like we're
>   Getting anywheres.
> It don't seem to me like we're getting anywheres at all."

Published just one month before the state of Alabama finally buckled and dropped all charges against Eugene Williams, Olen Montgomery, Willie Roberson, and Roy Wright, Boyle's "A Communication to Nancy Cunard" effectively captures the increasingly pessimistic mood of the Scottsboro defenders during the waning years of the decade.

By the verge of America's entry into World War II, the Scottsboro case had effectively disappeared as a rallying cry in American poetry—although a young African American poet, Robert Hayden, made one last valiant effort to resurrect it in his first collection of poetry, *Heart-Shape in the Dust*:

> America, America,
> listen to my song:
> a song of sorrow,
> a song of the slums,
> a song of the South,
> a song of Scottsboro,
> Scottsboro, Scottsboro,
> Scottsboro,
> with blood of the innocent
> upon its mouth. . . .
>
> Nine black boys
> condemned to die—
> and a million voices
> shouting: WHY?
>
> And a million voices
> rising, surging,
> beating like the sea:
> **O nine black boys,**
> **you shall be free,**
> **you shall be free. . . .**[78]

By this point, however, much as for Countee Cullen's "Scottsboro, Too" in the early 1930s, Hayden's "These Are My People" seemed somehow out of sync with the zeitgeist.

CHAPTER THREE

STAGING SCOTTSBORO

DURING THE 1930s, playwrights, too, were quick to exploit the dramatic potential of the vision of the Scottsboro case so forcefully articulated by the International Labor Defense, the Communist Party, and their allies—although their work was often filtered through the lenses of secondhand journalistic accounts and northern prejudices and stereotypes about the South. This was certainly the case with Myra Kay's agitprop drama *Scottsboro Boy*,[1] and Dennis Donoghue's *Legal Murder*, which opened at New York City's President Theatre in February 1934 for a mercifully brief run of eight performances.[2] In Donoghue's rendering of the case the nine Scottsboro defendants are initially depicted as happy-go-lucky neighbors who decide to hop a freight to Chicago to pursue their dreams of becoming radio singers. On the Chicago-bound train they are confronted by two white men, accompanied by two white girls, who order them to jump off the train. The boys rush the two men and knock them unconscious. The girls, described by one critic as "one of them a bum and the other an incipient bum,"[3] are unharmed. Shortly afterwards, a sheriff's posse raids the car. The tougher of the two women concocts a vivid story of sexual assault and her companions corroborate it. The rest of the play turns into a courtroom drama revolving around the trial of one of the boys, who, defended by a Jewish lawyer from New York, receives no justice from the judge, the prosecutor, or the jury.

Burdened with names like "Rastus Jackson" and "Dixie Mary," Donoghue's characters, one critic noted, spoke "in the traditional fashion of the vaudeville stage and no one appears quite real." This is a play, the critic stated, that "should not have come south of 125th Street."[4]

Reviewer W. A. Ucker pointed to *Legal Murder* as an "extra-judicial attempt to decide what judges and juries and millions of words have not yet been able finally to solve—the Scottsboro case," but these would be the kindest comments he would make. It is a "pretty bad" play, Ucker asserted, "blighted by a determination to depict the field hand Negro as a fellow who is constantly singing pseudo spirituals and blues songs at such times as he is not tap-dancing and feasting on fried chicken and other such fixin's." Even worse, Ucker continued, "every comic supplement Negro trait that was ever devised by men unfamiliar with the Negro of the deep South has been poured into it until those who should be shown as victims of terror emerge as court fools."[5]

Although *Legal Murder* incorporated many details of the actual Scottsboro case, Ucker noted, "these plays of propaganda are tricky things. They call for masters in their writing and production. They call for restraint, everlasting restraint and not the hot words of violent prejudice."[6] The lethal combination of technical ineptitude and flawed artistry, Ucker argued, consigned *Legal Murder* to the trash bin, and did not bode well for those dramatists who sought to transform the Scottsboro case into dramatic propaganda.

Appearing a mere six days later, John Wexley's critically acclaimed *They Shall Not Die* was by far a more gripping and successful dramatic production.

Born in New York City in 1907, Wexley began acting as a teenager and studied playwriting on a dramatic scholarship at the Washington Square Playhouse of New York University. He had already attracted public attention for his tough, prison drama *The Last Mile*, which opened on Broadway in 1930, introducing an unknown actor named Spencer Tracy.

Originally scheduled for a presentation at the National Theater in Washington, DC before its Broadway premiere, *They Shall Not Die* was abruptly canceled in Washington, leading to widespread rumors that racial considerations had contributed to the decision. Some people recalled that Marc Connolly's play *The Green Pastures* had been the occasion for many protests when it appeared in Washington, partly because a separate performance had been scheduled for African Americans, who were denied admission to regular performances; others remembered that the casting of the play had suffered because of District of Columbia laws preventing the appearance on the stage of any one under the age of eighteen. These minimum age laws were also cited as an impediment for the presentation of Wexley's play, which included three child actors in the cast—although it did little to ease public suspicions about the racial motives underlying the decision.

Based upon the sequence of events from the arrest of the Scottsboro Boys in 1931 through the second trial of Haywood Patterson in Decatur, Alabama in 1933, *They Shall Not Die* was produced by the Theatre Guild at New York City's Royale Palace, opening on February 21, 1934.

Sprawling and ambitious, the play featured a cast of more than eighty, including Tom Ewell, Ruth Gordon, Dean Jagger, and Claude Rains, and elaborate sets that ranged from the jail in "Cookesville" to the home of Lucy Wells (Ruby Bates) in "Humbolt" to the New York City law offices of Nathan G. Rubin to the courtroom in "Dexter." Wexley generally stayed close to the facts of the Scottsboro case as they had been presented to the public. As he recalled many years later:

> I studied the court records and interviewed people. I went to see the lawyer who defended the boys and followed him to trials, observing his style. Oh,

Figure 3.1 Dust jacket of John Wexley's *They Shall Not Die*. Source: Syracuse University Library.

and I met the mothers of the boys, and the key witness, a prostitute, who gave evidence that she had been instructed to lie. Ruby Bates was her name. She'd come North to raise funds for the defense.[7]

When the public record did not exist, however, Wexley gave free rein to his imagination. He deliberately applied a gossamer-thin veneer over the people and places involved in the Scottsboro case—Victoria Price was renamed Virginia Ross, Haywood Patterson became Heywood Parsons, the International Labor Defense became the National Labor Defense, the NAACP was renamed the American Society for the Progress of Colored Persons, and so on. There would be no confusion among 1930s audiences about who these figures and organizations were. Nor would there have been any doubt about where Wexley's political sympathies rested, for he clearly advocated the perspective of the International Labor Defense.

Act I of *They Shall Not Die* is a dramatic reconstruction of the Scottsboro case, from the March 1931 arrest of the Scottsboro Boys through the Decatur trials of 1933. In the jail in Cookesville, through the contrapuntal dialogues of two white deputy sheriffs, on the one hand, and three white prisoners, on the other, Wexley skillfully conveys the boredom, the economic plight, and the social and racial tensions of "Cookesville" during the early years of the Depression; there are already broad hints, in the snatches of dialogue, about the measures some white women have taken to ease their economic burdens. Against this backdrop, the first news trickles in of a fight between black and white hoboes on a freight train from Chattanooga and of the violent expulsion of many of the whites from the train. This report is immediately followed by an account of the discovery of two white women still on board the train. The participants in this event arrive in successive stages: six white male hoboes, the two white women, and the nine accused black youths.

In the ensuing chaos, a story begins to be constructed. In his first interrogation of Virginia Ross and Lucy Wells, Sheriff Trent seems inclined to regard them as nothing more than traveling prostitutes who have freely bestowed their favors on the train. Once Virginia and Lucy are out of his sight, however, Sheriff Trent begins an intense conversation with County Solicitor Mason, who—up until this point—has been considerably less absorbed by the situation than Trent:

> MASON: All right. What's on your mind, Sheriff?
>
> TRENT (*with suppressed anger*): Plenty! Them tramp whores have been crossin' the State line and doin' business on that train with these white hoboes. . . .
>
> MASON: And the niggers . . . ?

TRENT: The niggers?

MASON: Yes . . . ?

TRENT (*with increasing anger*): . . . them black bastards seen 'em gals and got themselves together, outpopulated an' beat up them hobo kids and threw 'em off the train. Then the niggers jumped the gals an' . . .

MASON (*with a cautioning gesture*): Just hold on, Trent. I want to get this straight. These heah girls don't look like to me they been attacked.

TRENT (*with amazement and somewhat hurt*): Whut you tryin' tuh say to me Luther . . . ?

MASON (*firmly*): I'm trying to say this. . . . If these girls had been assaulted against their will, they wouldn't be acting the way they are. They would be crying all over the place. They would be all hysterical and nervous. Their clothes would be torn. . . .

TRENT (*angered, annoyed*): You ain't sayin' Luther, that them niggers were left alone with these white gals and didn't try to . . . ?

MASON (*interrupting with some scorn*): No! They didn't need to try. These whores just took them on for whatever they could get. . . .

TRENT: Luther, you ain't goin' to let them black bastards get away with something like that?

MASON: No. . . . I'm not letting them get away. . . .

(*HE seems to be listening to Trent only with one ear and to be thinking of a plan of procedure.*)[8]

Trent finally makes what appears to be the decisive argument:

TRENT (*in a rage*): I don't keer if they are whores . . . they're white women! You think I'm gonna let them stinkin' nigger lice get away from me? Like hell I am! They're gonna git whut's comin' to 'em long as I'm the law around here. . . . (*He is at the height of his temper and his feelings run away with him.*) What the hell will folks heah say of us . . . ? Why they'll spit on us if we don't git them niggers when we got the chance. . . . The hull county, the hull State, the hull South'll be down on our haids.[9]

Thus Trent effectively establishes the terms of the narrative that will be told and the bargain is sealed through a series of exchanges with Victoria

Ross, who has established herself as the chief negotiator for the two white women:

> MASON (*in the interim. Rather friendly*): You understand, Virginia. . . . We just can't let these niggers get away with such things because of the bad effect on other niggers.

> VIRGINIA (*grateful that they are treating her as an equal*): Well, I guess you're right. . . .

> MASON (*as Trent enters*): After all, if we let 'em git away with this once, a white lady wouldn't be safe any more.

> VIRGINIA: Yeah. . . . they git uppity mo' an' mo'. . . .

> TRENT: Sho'. All kin' o' fool notions nowdays. . . . some even talkin' of votin' an' down in the Birmingham steel . . . they're havin' all sorts o' trouble with 'em . . . and down 'round Tallapoosa I heerd they're formin' a share-croppers' union. . . . Did you know that, Luther?

> MASON (*ignoring his question*): I know how you feel, Mrs. Ross. I know you're ashamed. It's not pleasant to have everybody know of such a disgrace. I know too well how you must have suffered. But you must realize too what a splendid brave thing you'll be doing for our kind of people. . . . And have no fear, this state and Hatcha-chubbee County will not soon forget your sacrifice. . . . In fact the whole South. . . .

> VIRGINIA (*impressed but cautious*): Well . . . I don't keer to git in no trouble. . . .

> MASON: You're certainly headed for plenty of that if you let folks get the idea you took on those niggers of your own free will. . . .

> VIRGINIA (*mechanically denying*): I never tuk none of 'em niggers on. . . .

> MASON (*meaningfully*): But you heard the doctor tell what he found in you. . . . Yes. . . . it means a great inner struggle but you are only the victim of a cruel fate. . . . no one will blame you for telling the truth. For having the courage to tell the truth. The newspapers, the Governor, every man, woman and child will thank you and praise you. The whole state will have your name on its lips. Your picture will be in every newspaper. . . .

> TRENT: Sho'. . . . instead of sayin'. . . . what a low trash. She wouldn't even help the law and admit what the niggers done tuh her . . .

MASON: And then losing your job too. . . .

VIRGINIA: Yuh sho' I wouldn't lose my job if I tell. . . .

MASON: Yo can hold me personally responsible, Mrs. Ross. Sheriff Trent's a witness.

VIRGINIA (*shrewd, smiling*): Well, couldn't I have a new dress fo' tuh take them newspaper pictures in? This doesn't look so good. . . .

MASON (*smiles*): Why certainly, we can arrange that. Most of the women here would be proud to help out.

VIRGINIA: I would 'preciate it. . . . Lucy too. . . . I'm sho'. An . . . er . . . maybe a li'l change too? Jest a coupla dollars fo' the time I have tuh stay here. Yuh see I have an old sick maw who I support. . . .

TRENT (*becoming annoyed*): Now looka heah, gal. . . .[10]

Through carefully constructed scenes such as these Wexley makes a plausible case for the behind-the-scenes maneuvering, badgering, and outright intimidation that could have led Virginia Ross/Victoria Price to testify the way she did during the Scottsboro trials. As we shall see momentarily, Wexley insists upon drawing a sharp distinction between the character of Virginia Ross, whom he depicts as a shrewd, worldly perjurer, and that of Lucy Wells, who is presented in terms of guilt-stricken innocence.

Another major line of development in the play revolves around the moral and spiritual regeneration of Lucy Wells, the more uncertain and troubled of the two white women accusers. In act II, scene 1, set several weeks after the trial in Cookesville, we witness Lucy back at home in her squalid surroundings in the "nigger-town" section of Humbolt. She is clearly depressed, still works intermittently as a prostitute, and is slowly realizing that the terms of her life have not changed at all—as the following exchange with her mother makes clear:

LUCY (*still sullen and rather introspective*): I'm feeling I didn't git hardly enough fo' whut I done fo' 'em. Maybe I shouldn't never've done it at all.

MRS. WELLS: Whut yuh talkin', gal? Yuh jest had tuh do right by the law. Whut would decent people say of us if yuh hadn't? Lo'd knows they look down at usaplenty as is . . . an' if yuh . . .

LUCY (*with some feeling*): Well, they still look down at us. They do. They promised me all so'ts of things. They promised me steady work at the mill . . . an' heah they haven't given me mo' than seven or eight days in all the weeks sence the trial. . . .

Figure 3.2 Victoria Price. Source: Files Crenshaw, Jr. and Kenneth A. Miller, *Scottts-boro: The Firebrand of Communism* (Montgomery, AL: Brown Printing Company, 1936).

MRS. WELLS (*speculatively*): Well, maybe things is slow. . . . but they'll pick up after a while an' . . .

LUCY: Well, 't ain't so. Things is slow . . . but Virginia Ross . . . she's gittin' work. . . .

MRS. WELLS (*surprised*): She is?

LUCY: Sho' she is. She's gittin fo' an' five days a week an' I'm only gittin' two days every second week. . . . (*Flaring up*) An whut's mo' she goes astruttin' 'round town like a fightin' cock, talkin' how smart she is. . . . an' how she showed up at Cookesville an' how she had tuh shet me up . . . an' how stupid and dumb I was all the time. . . .[11]

Lucy shyly reveals to her mother that she has met a traveling salesman, Russell Evans, at the local drugstore. The two are clearly attracted to each other. Evans apparently sees something deeper in Lucy than her present circumstances and appeals to her to "try to start a sort of new life and jest fo'get your past." It is only then that he realizes that Lucy has been involved with the Cookesville trial; he leaves town shortly afterwards.

In *They Shall Not Die*, the plight of the nine Scottsboro defendants provides the occasion for the play, but their lives are secondary to the dramatic action. Wexley provides graphic scenes of their brutal interrogation in the Cookesville jail; and with deft dramatic strokes he allows his audience to distinguish among Roberts (Willie Roberson), Ozie Purcell (Ozie Powell), Gene Waters (Eugene Williams), Olen Moore (Olen Montgomery), Heywood Parsons (Haywood Patterson), Roy and Andy Wood (Roy and Andy Wright), Morris (Clarence Norris), and Warner (Charlie Weems). With the exception of Heywood Parsons/Haywood Patterson—distinguished in life and art by his truculent, defiant outlook—the Scottsboro Boys generally function as recipients of the actions of others, whether it is their jailor-torturers or their lawyers. Unlike Langston Hughes in *Scottsboro Limited*, who viewed the Scottsboro Boys in heroic terms, Wexley offers a spare, unsentimental view of them.

In act II, scene 2, we encounter the Scottsboro Boys in the "negro death cell in Pembroke Prison" in the aftermath of their first conviction. Here Wexley reenacts the struggle between the International Labor Defense and the NAACP over who will claim the right to defend the Scottsboro Boys. In the presence of the prison keeper; a black preacher; Lowery, a white lawyer from Birmingham; and Treadwell, a representative of the American Society for the Progress of Colored Persons, the boys fawn, fret, and waver before the majority sign an agreement permitting the A.S.P.C.P. to represent them in their appeal. In the meantime, the play exposes the utter bankruptcy of the black clergy and its complicity with the A.S.P.C.P.:

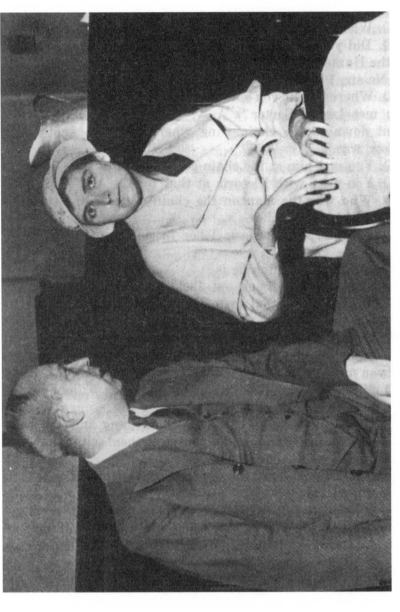

Figure 3.3 Samuel S. Leibowitz and Ruby Bates. Source: Files Crenshaw, Jr. and Kenneth A. Miller, *Scottsboro: The Firebrand of Communism* (Montgomery, AL: Brown Printing Company, 1936).

Figure 3.4 "Scottsboro Mothers." Left to right: Ruby Bates, Mamie Williams, Viola Montgomery, D.C. civic leader Julia West Hamilton, Janie Patterson, and Ida Norris. Photograph by Addison Scurlock, Washington, D.C., May 1934 (Smithsonian Institution: National Museum of American History).

> PREACHER (*takes from pocket a prayer book and speaks to them. After a few words he begins unconsciously to chant rhythmically*): My chillun! I want tuh put the Lo'd in yuh. I want yuh tuh feel that the Lo'd Almighty is in us an' is in the great A.S.P.C.P. . . . This N.L.D. is a contraption of the devil's an' Satan. He sent them tuh make trouble an' bring down hate an' prejudice on God's colored chillun. . . . if yuh all is 'lectrocuted an' dies yuh'll all go tuh Heaven sho' as yuh're born if yuh're sho' yuh ain't had a hand in this terrible crime. . . . An' so I bless yuh and warn yuh tuh fergit that N.L.D. devil's bunch an' sign up with the blessed A.S.P.C.P. Oh Lo'd, looka down on these po' misguided nigra chillun an' lead 'em safe an' holy tuh yo' kin'ly light. . . .[12]

The atmosphere dramatically shifts with the arrival of the National Labor Defense team, led by Rokoff (Joseph Brodsky) and accompanied by Attorney General Cheney (George Chamlee). Without any apparent regard for the presence of the prison keeper, Rokoff presents a straightforward and unequivocal view of the mission of the N.L.D.:

> What can the N.L.D. do? I'll tell you what we can do. First, we'll get the finest lawyers in this country to fight the courts at their own game . . . but more important than that, we'll go to the workers of America, to the workers of the world. We have proof that you're innocent of these rape charges. We'll show them this proof. Then we'll say to them: Black and white workers of the world! Workers of America! Down in the South nine innocent boys are being put to death because they have black skins. Are you going to stand for that? And they will answer with a shout that will ring around the whole world. . . . NO. We will not. Yes, we will force the South and those in the South who are trying to murder you . . . we will force them to free you. Yes, they will. They'll be afraid to keep you, afraid to kill you . . . They'll be afraid of fifteen million black workers who will stand shoulder to shoulder with fifty million white workers and who will roar. . . . Don't you touch those boys! Don't you dare touch those black children workers . . . ![13]

Rokoff's rousing speech is powerful enough to persuade Parsons (Haywood Patterson), whose own powerful words turn the tide in favor of the N.L.D.:

> listen tuh me, you niggers! Yuh all purty dumb. Maybe yuh don't understan' his talk. But it 'peared tuh me he was talkin' our own language an' I understood ev'ry word he say. An' he say a-plenty! He ain't no yaller-belly tuh sell us out. Lo'd A-mighty . . . when he talked I felt jest as strong as a bull. I felt I could bust open these heah bars. An' I'm a-tellin' yuh all dat I don't keer if Gawd or the debbil or the N.L.D. saves me, I wanna be saved. An' this heah man kin do dat. . . . Yes right down heah in the South. So I

say tuh yuh all . . . Sign up! Sign up, niggers, befo' he gits angry an' changes his min' wid us dumb bastards.[14]

Act II, scene 3 returns to Lucy Wells's home "many months later." Lucy has clearly continued to suffer—physically, emotionally, and spiritually. Having heard that Evans, the traveling salesman, is back in town, she has written a letter to him, admitting the truth about the "Cookesville" affair and her role in it, and entrusted its delivery to a local black youth. Under close questioning from her mother about the letter, Lucy falls apart and confesses:

> LUCY (*she speaks as if what she has to say could not be held back a moment longer. It pours out of her*): I . . . I cain't sleep nights. That's why I got so sick. I'm all run down with thinkin' of it. Thinkin' of them po' nigger kids, goin' tuh burn any day on that 'llectric chair. I dream. . . . I dream of them screamin' an' yellin' in pain. . . . I see myself, lightin' fires an' helpin' tuh burn them. . . . (*She sobs bitterly.*)[15]

For Wexley, writing in 1933–34, the character of Lucy Wells is the crux of the drama, just as the testimony of the actual Ruby Bates is, presumably, the key to the resolution of the Scottsboro case. Lucy Wells is arguably one of the most significant characters in *They Shall Not Die*, and Wexley invests considerable attention in charting her moral regeneration.

Lucy's letter falls into the hands of the local authorities, who, with the backing of the county solicitor Luther Mason, bully her into recanting her statement. Evans arrives immediately after the authorities leave. In rapid succession, Lucy tells him everything; he is moved by her simplicity and sincerity, they vow their love for each other, and flee town to evade the authorities. This scene sets the stage for the powerful courtroom exchanges that conclude the play—paving the way for Lucy's dramatic appearance as a witness for the defense.

The final scene of *They Shall Not Die* is cast in terms of an epic struggle between the forces of good and evil, with Wexley presenting the jurors and southern whites in images strongly reminiscent of the rhetoric of Samuel Leibowitz and Muriel Rukeyser—although he seems to harbor a fondness for the presiding judge (Horton):

> *The jurors and white audience are for the most part . . . lean, hard-faced, thin-lipped individuals, raw-boned and provincial. Many wear heavy, mud-caked, cowhide boots and overalls. The Judge, ironically enough, resembles Abraham Lincoln without the beard. He speaks in a soft drawl . . .*[16]

In the final scene of the play, Wexley follows the historical court transcript fairly closely, allowing the often outrageous, verging upon the surreal, interactions between the defense, represented by Rubin, and the state, represented by Dade (Thomas Knight), to speak for themselves. The scene recapitulates the state's argument for the routine exclusion of African American jurors from trials in the region and the state—in bold defiance of the Thirteenth, Fourteenth, and Fifteenth amendments to the U.S. Constitution—while the table of northern reporters functions, in effect, as a Greek chorus registering its amazement and dismay at the proceedings. Virginia Ross's lurid testimony seems to collapse in the face of the obvious facts, but Wexley's script exposes the relentless determination of the state to press its case in spite of the compelling arguments to the contrary.

Wexley is particularly graphic in his depiction of how Rubin's egalitarian instincts constantly clash with deeply entrenched southern racial mores, and in his dramatization of the bedrock anti-northern hatreds and anti-Semitism that persistently rise up to challenge—and ultimately defeat—Rubin's best efforts. In spite of Parson's emphatic denial of the charges, Doctor Morton's fairly straightforward testimony suggesting the improbability that the two women had been raped, Lewis Collins's account of the evening he spent with Virginia Ross and Lucy Wells the night before the arrest of the Scottsboro Boys, and Lucy Wells's surprise appearance and dramatic repudiation of her previous testimony, the outcome of the case clearly hinges upon the ability of the state to appeal to the prejudices of the jurors.

Circuit Solicitor Slade begins the summation of the state's argument:

> SLADE (*strides to the front of the jury. He commences almost at a climax and somehow keeps it up by virtue of his tremendous physical power. He almost roars his words as he stands red-faced, bull-necked, a huge six foot mass of beef*): Yo' Honor and gentlemen of the jury and my friends of Dexter. First, I would like tuh review the evidence fo' you. You jest saw this Lewis Collins, or better *Collinsky*, with his New York clothes. Why, my friends, jest two mo' weeks with this Rokoff an' he'd a ben down heah with a pack on his back atryin' tuh peddle us goods. Are yuh goin' tuh stand fo' this so't of thing?. . . .
>
> The prettiest Jew yuh ever seen, this Lewis Collins, amovin' his hands thisaway and thataway. . . . They think they kin come down heah tuh obstruct justice in this heah co'troom. Yes, it was the N.L.D. who brought in Lucy Wells, an' bought her soul. The same N.L.D. who put them fancy clothes on Collins, New York City clothes. And I tell yih, gentlemen, this Lucy Wells is guilty of perjury right heah in this co't. And theah is such a thing too, as subornation of perjury. . . .
>
> (*Points again at defense.*) That Wells gal couldn't tell yuh all the things that happened in New York, 'cause part of it was in the Jew language. . . .[17]

State attorney Dade delivers the *coup de grace*: "Gentlemen of the jury, tell 'em, tell 'em that Southern justice cannot be bought an' sold with Jew money from New York."[18]

Rubin's detailed and impassioned summation of the case for the defense is quickly overshadowed by Dade's strident appeals for the summary execution of the Scottsboro Boys. Even Rubin, however, seems unprepared for the denouement of the trial:

> (*The Jury stands, turns and crosses to the jury-room, in silence. The court-audience is silent too. The Guard shuts the door on them. A pause. Silence. Then suddenly from the jury-room, a sound of loud laughter, raucous and derisive. As he hears this, Rubin is startled for a brief instant and turns slowly, not knowing where the sound is coming from. Then with a half-audible sound and an expression of mixed astonishment and dismay, he rises slowly and speaks.*)

> RUBIN: If the court please . . . I have seen and heard of many strange and crazy things my time, but I have never heard of anything like that . . . in there. (*He gestures toward the jury-room.*) But I'm not through yet. Let them laugh their heads off . . . this case isn't finished yet. . . .[19]

Although the curtain rings down with the jury still deliberating, Wexley clearly confers the moral victory upon Rokoff and Rubin, in the spirit of classic agitprop endings—but minus the rhetorical flourishes and apocalyptic vision of, say, Hughes's *Scottsboro Limited*:

> ROKOFF (*rises and stands at RUBIN'S side*): No . . . and our fight isn't ended either . . .

> JUDGE (*rapping his gavel*): This . . . this is out of order. . . .

> ROKOFF (*continuing over the interruption*): You have the jurisdiction to stop us in this court . . . but there are hundred of thousands of men and women meeting in a thousand cities of the world in mass protest against the oppression and ownership of man by man. . . . and over them you have no jurisdiction. . . .

> RUBIN (*inspired and fired by ROKOFF*): No . . . we're not finished. We're only beginning. I don't care how many times you try to kill this Negro boy. . . . I'll go with Joe Rokoff to the Supreme Court up in Washington and back here again, and Washington and back again. . . . if I have to do it in a wheel-chair . . . and if I do nothing else in my life, I'll make the fair name of this state stink to high heaven with its lynch justice. . . . *these boys, they shall not die!*

> (*Laughter from the jury-room dies down and the court-audience stare at him with eyes and mouths agape. . . .*)[20]

In the midst of Broadway's economic woes in the mid-1930s, *They Shall Not Die* had a fair run of over sixty performances, attracting along the way the critical attention of both the mainstream and left-wing press. Virtually all of its reviewers noted its anger and raw power. F. Raymond Daniell, who had reported on the Scottsboro case for the *New York Times*, praised the play for its realism and accuracy.[21] Other critics concurred about its factual accuracy. Writing for the *New York Evening Journal*, John Anderson commented: "Most of the anger and ugliness and some of the pity that go to make up that celebrated Scottsboro case have been lumped into a play, a furious and bitter play. . . . It is," he continued, "apparently, a closely truthful story of that trial, which already takes its place with the Mooney and Sacco-Vanzetti cases as a rallying point against legal oppression."[22]

Arthur Ruhl adopted a more detached and clinical perspective toward the play, emphasizing its status as propaganda and pontificating:

> The artistic horizon, or "ceiling," as to say, of any such piece is, in the nature of things, sharply limited. As the spectator knows beforehand the general nature of the story, the usual element of surprise is more or less eliminated. . . . In short, the evening becomes not so much a play as a sort of "rally,"—most of the spectators coming to have corroborated and strengthened opinions already made up.[23]

In spite of his apparent disdain for dramatic propaganda, Ruhl conceded that "The Negro parts were exceedingly well-cast—by boys, that is to say, or Negroes who looked like the bewildered, utterly ignorant and simple young blacks which the author would have to believe the defendants are."[24] Still, he warned his readers, "The evening is interesting, but uneven in quality, and its frankly controversial and occasionally painful character should be understood by those who go to the theater looking merely for entertainment."[25]

Robert Garland was so taken with *They Shall Not Die* that he devoted a second column in the *New York World Telegram* to a consideration of the implications of Wexley's "strident, rambling, disturbing play. . . . Not in all my theatergoing days have I seen such righteous indignation in the theatre. A righteous indignation that, like a spark along a fuse, runs from play to player, from player to audience."[26]

Garland took exception, however, to Wexley's depiction of the fundamental frame-up that triggers off the action of the play, arguing that it "weakens the structure of the play. . . . It is never quite on the level. It has no actual motivation." Speaking confidently from a northern, liberal perspective, Garland freely confessed his skepticism about how Scottsboro occurred:

> Once you admit that nine Negro boys are framed for no reason save that they are Negroes—and Negroes with all the cap-tipping deference you might expect from dear old Uncle Tom—the remainder of Mr. Wexley's

play is understandable. But, along with Brooks Atkinson, I find much of the drama's groundwork open to dispute.

Between you and me, I doubt if Southern minions of the law lie endlessly in wait to pin some imagined misdemeanor on some frightened colored person.

Once that frightened colored person has committed something the unfrightened white man believes to be a crime, I, along with Mr. Wexley, say "Heaven help him." But . . . I am not yet convinced that each and every Southerner, from Governor to turnkey, hides constantly behind an electric chair with the hatred of a Negro in his heart and a frame-up of a Negro in his mind.[27]

Like Scottsboro counsel Samuel J. Leibowitz, Garland apparently badly misread the conditions that led to the eruption of the Scottsboro case, blithely ascribing the possibilities of reason to social and political circumstances where it was a rare commodity.

The Communist Party press hailed *They Shall Not Die*, albeit with some political reservations. Writing in the *Labor Defender*, Isidor Schneider praised Wexley for his "superb use of the Scottsboro case," pointing particularly to how in his selection of the details of the play "Mr Wexley has been able to emphasize one of the two outstanding lessons of the Scottsboro case with tremendous clarity and vigor":

> The scene in the jail where the sheriff and the solicitor work out the rape charges and lay the basis for the frame up; and later, the court scene where the state's attorneys with the outright connivance of the judge, play upon the race prejudice of the jurors and of the mob in the courtroom, these show in action and with unequalled dramatic force, the white ruling class at its vicious work of preserving the division between the white and black masses, the division which preserves the ruling class in its privileges and power.[28]

Schneider also praised Wexley for his vivid dissection of the bankruptcy of the Negro reformist organization (NAACP) when it attempts to persuade the boys to place their case in its hands:

> The role of these reformists' organizations of the white ruling class is made unforgettably clear. The value of the play in demonstrating these two factors of the class struggle in the South, the deliberate use of race conflict by the white ruling class of the South and the alliance of the Negro bourgeoisie with that ruling class, is incalculable.[29]

Still, Schneider pointed out,

> the most important sector of the struggle was neglected. I refer to the mass pressure, the omission of which was almost half the battle. . . . Had it been

told directly, the emphasis would have been shifted to balance the frame up artists of the white ruling class. There would have been shown the national secretary of the International Labor Defense, a negligible figure in the play and other leaders in their wonderful work in organizing mass action; and to balance the hate blinded southern white mob poisoned in mind by the boss class, there would have been shown the truth led masses, Negro and white mobilized by our International Labor Defense against the enemies of the working class.[30]

Harold Edgar informed the readers of the *Daily Worker* that *They Shall Not Die* was "a play to be seen, to be applauded, even to be cheered." At the same time, however, he pointed out that Wexley had given insufficient attention to the economic determinants of the Scottsboro case, an issue only hinted at in Rokoff's encounter with the boys.[31] But Edgar reserved his harshest judgments not for the play, but for its audience:

> A large part of the Theatre Guild subscribers and the so-called "general public" which makes up Broadway "houses" are simply miscast in relation to this play: they will not act as an audience should at such an event. They try as hard as possible not to be "taken in": their whole idea of "art" is something that "moves" you without disturbing you, something that permits you to shed a consoling tear and which one "raves" about and forgets after one has seen the next big hit. If they take sides, if they should forget too completely that what they are seeing on the stage is truly an image of life and involves a responsibility on their part, their "artistic" pleasure might be spoiled. They do not want to be involved: they prefer to sit contemplatively in their seats and regard the spectacle solely as a "show."[32]

This weakness could be easily remedied, Edgar maintained, by readers of the *Daily Worker* becoming part of the audience for *They Shall Not Die*, transforming the theater, and teaching "the Theatre Guild's usual audience and the dramatic reporters that the world of the theatre is not bounded by Eugene O'Neill on the north and by romantic masquerades on the south."[33]

While Michael Gold—widely regarded as the cultural commissar of the literary Left—praised the Theatre Guild for bringing the working class into "that hitherto stuffy museum of bourgeois decay which was the Theatre Guild,"[34] he also took exception to the considerable attention Wexley gave to the flamboyant, middle-class defense counsel Rubin, "a most serious political blunder" that distracted the audience's attention from the mass demonstrations and telegram campaigns the International Labor Defense and the Communist Party had organized to mobilize public opinion about the case.

They Shall Not Die clearly struck a powerful chord among theatergoers, attracting on March 4, 1934 more than one thousand subscribers to the Theatre Guild to a symposium on the play and on the Scottsboro case on which it was based.[35] Scottsboro defense counsels Samuel Leibowitz and Joseph Brodsky participated in the symposium, sharing the stage with John Wexley; William Patterson, representing the International Labor Defense; Ruby Bates, one of the Scottsboro Boys' original accusers; the playwright and critic Elmer Rice; the writer and critic Ernest Boyd; Philip Moeller, the director of the play; and others. The symposium was largely devoted to a spirited discussion of the social implications of the Scottsboro case, but the *New York Times* reported at least one moment of comic relief:

A few seconds after Mr. Brodsky had finished his address his place was taken by his stage counterpart—Rokoff (otherwise Louis John Latzer)—whose speech and stage mannerisms were so like Mr. Brodsky's that the attorney seemed unable to decide whether to be amused or dismayed.[36]

Ernest Boyd offered a sharp public critique of Wexley's play, while Elmer Rice reminded the audience of the significance of the Scottsboro case in global terms: "We have tended to be pretty critical of the Hitler regime, of Mussolini, even of Russia, but right at our doorstep lies something which overshadows anything being done in Europe."[37]

Among the regular theater audience, *They Shall Not Die* was powerful enough to elicit the following letter to Alabama Governor Bibb Graves from a fellow citizen:

The play "They Shall Not Die" is a direct reproduction of the Scottsboro trial. Its effect upon injustice in the South will be the same as that of "Uncle Tom's Cabin" on slavery.

The fact that Alabama is being truthfully paraded before the world in this case makes it very embarrassing, intimidating and uncomfortable for an Alabamian who travels elsewhere.

Everybody the world over now knows that those nine negro boys are innocent and were framed so why continue with such a "farce of a trial?" Why make Alabama appear as the jackass of the South for justice?

This play is a howling success and the North, East and West are raving to see it. Mass action is being consolidated and united as never before and it will be impossible to kill those nine innocent boys.

Why not get the best minds of Alabama together and find the best way out of this mess? There is no doubt "They Shall Not Die" has put Alabama on the spot and something must be done and that right early.

P.S. I was born and bred in Alabama and love that State. That's why I write this.[38]

From the perspective of doctrinaire Communist Party critics, a far more ideologically satisfying play was Paul Peters and George Sklar's *Stevedore*, which opened at New York City's Civic Repertory Theatre on West 14th Street on

April 18, 1934—two months after *They Shall Not Die*; the play ran initially for 111 performances, then, after a brief hiatus, resumed production on October 1, 1934 for another run of 64 performances.

Paul Peters (the pseudonym of Harbor Allen) was born in New York City and educated at the University of Wisconsin. An active Communist, he was a member of the staff of *New Masses*—which had published his one-act play *Mr. God Is Not In* in December 1926—and later served on the editorial board of the International Labor Defense's magazine *Labor Defender*. George Sklar, born in New York City in 1908 and educated at Yale University and the Yale School of Drama, was a committed Marxist and widely regarded as "one of the leading Marxist playwrights"[39] of his generation. As members of the Executive Board of the Theater Union, Peters and Sklar subscribed to the manifesto of the company:

> We produce plays that deal boldly with the deep-going social conflicts, the economic, emotional and cultural problems that confront the majority of the people. Our plays speak directly to this majority, whose lives usually are caricatured or ignored on the stage. We do not expect that these plays will fall into accepted social patterns. This is a new kind of professional theatre, based on the interests and hopes of the great mass of working people.[40]

In their collaboration on the production of *Stevedore*, Peters and Sklar drew upon a play that Peters had originally written about black longshoremen in New Orleans—sections of which he had published as *Wharf Nigger* and *On the Wharf* in the November 1929 and April 1930 issues of *New Masses*.[41] In the original version of the play, the New Orleans setting served as a springboard for Peters's dramatic ruminations about outbreaks of racial violence in Chicago and St. Louis. Working closely with George Sklar, however, Peters broadened the play's scope to include the challenge of organizing labor unions across racial lines. As he observed in an interview published in the *Daily Worker*, the original version of the play was unsuitable for the Theater Union "because . . . it dealt with the Negro problem from the race angle rather than from a Marxist approach."[42] In its refurbished form, allusions to the Scottsboro case quickly moved to the background.

The first scene of *Stevedore* presents the audience with the event that provides the critical backdrop for the play: the false accusation of rape by a white woman of questionable morals. One evening a white man assaults his former lover, Florrie Reynolds, in the back yard of her house in New Orleans after an angry exchange with her. He flees, and when Florrie's husband rushes out of the house in response to her screams, the following exchange occurs:

> FLORRIE (*whimpering*): I was taking out the garbage.
>
> FREDDIE: What happened, Florrie?

FLORRIE: Somebody just jumped out of the dark and grabbed me.

NEIGHBORS: Good God!
        Did you see who it was?
        Who was it, Mrs. Reynolds?

FLORRIE: It was—it was—(*She breaks into sobs.*)

FREDDIE: Who was it? Who did it?

FLORRIE: It was—(*weeping desperately*)—a nigger!

BLACK OUT[43]

In the hands of Peters and Sklar, the false accusation of rape is depicted primarily as a means of regulating and disciplining the black community, particularly socially and politically assertive black men. Unlike John Wexley, who—through his exploration of the character of Lucy Wells in *They Shall Not Die*—devotes considerable attention to the complex interaction of race, economics, status anxiety, and sexual mythology in the Jim Crow South, Peters and Sklar are relatively indifferent to Florrie Reynolds as a character; after fulfilling her function as the springboard for the action of the play, she disappears after act I, scene 2.

Florrie's accusations provide the occasion for a police roundup of "all suspicious Negroes in neighborhood," including Lonnie Thompson, a worker at Oceanic Stevedore Company, who distinguishes himself by his refusal to follow orders in the police lineup.

One key site in *Stevedore* is Binnie's lunchroom near the dock, staffed by Binnie, Uncle Cato, and Lonnie's girlfriend, Ruby. This is where the black workers eat, talk, argue, and give vent to their frustrations and dreams. Together they represent a spectrum of African American strategies for coping with racial oppression, with Lonnie representing the most militant response—as is clear in the following exchange:

> **LONNIE:** God damn dem, anyhow. What dey think I am? Do I look like some kind of animal? Do I look like somebody who'd jump over a back fence and rape a woman?
>
> **BINNIE:** Sit down and eat yo' beans, boy.
>
> **LONNIE:** What right dey got to treat us dat way? Line up, nigger. Step out, nigger. Get back in yo' cage, nigger!
>
> **SAM:** Take it easy, boy. Take it easy.
>
> **LONNIE:** I think of dat, Sam, and I just burn up inside. Dey don't treat us like human folks, Sam. We just dogs to kick around.

SAM: You better forget it boy. Forget it.

LONNIE: Forget it!

CATO: Dat's what you better do, son. You talk too much. Don't pay to talk too much. Nassuh.

LONNIE: You got fire inside of you, you got to spit it out.

BINNIE: You better watch where you spitting, boy.[44]

Much like John Wexley's Heywood Parsons, Lonnie Thompson commands our attention because of his attitude of uncompromising resistance towards the racial status quo. It turns out that Lonnie has been "carrying on" with Lem Morris, a white union organizer working on the docks who is known to be a "Red." When his fellow black workers complain about being cheated on their paychecks, Lonnie organizes a delegation to confront the company about their exact wages. Following an angry exchange with Lonnie, Walcott, the superintendent, knocks him down. After Lonnie, restrained by a coworker, leaves, Walcott explodes:

> That god damn shine's been hanging around with Reds, that's where he gets those notions! Organizing! Equality! Christ, those coons are just running riot! This country won't be fit to live in any more. Look at all the rapes they've been pulling in this town. Look at what they did to that Reynolds woman. And did they catch anybody for it? No! Where the hell are the cops? By God! I'm going to call up Severson. (*He dials with angry motions as he talks.*) If they don't do something about it, I'll organize a posse and go after them myself. It's high time those niggers were put in their place.[45]

From this point on, Lonnie Thompson is clearly a marked man. Significantly, the police search for Florrie Reynolds' "assailant" escalates in the aftermath of the altercation between Walcott and Lonnie. The ensuing tension directly contributes to the assault by white thugs on a meeting Lonnie has organized to discuss a union with white workers. At a critical moment in this scene when Lonnie, bruised and cut by his white attackers, seems to be on the verge of rallying his comrades to strike back against their attackers, he is arrested as a suspect in the Reynolds case. Without Lonnie, the workers revert to their customary passivity and servility in the face of white authority:

> (*The men on the wharf surge forward as if to help Lonnie. They stop as they become aware of Walcott, who turns and looks at them. There is a tense silence. Walcott starts moving slowly across the wharf. As he approaches them, they return sullenly to work. The grinding of a winch is heard from the wharf apron . . .*)[46]

At the beginning of act II we are introduced to Lem Morris, "a simple, rugged white longshoreman of about forty," in the Union Hall, planning a meeting of black and white dockworkers later that evening. Word arrives that Lonnie has been arrested—and that he has escaped. Lem immediately begins to mobilize the workers for Lonnie's defense because "he's a member of the union and we've got to help him get away."

Al, a white worker, takes sharp exception to any attempt to help "a rape nigger." Lem systematically dismantles his argument:

> LEM: Rape nigger. Where do you get that noise?
>
> AL: That's what they arrested him for, didn't they?
>
> LEM: You mean, that's what they framed him for.
>
> AL: Aw nuts. You're always yelling about frame-ups. How do you know he didn't rape her?
>
> LEM: Listen, you know damned well, the only reason they arrested him was to keep him from organizing the Stuyvesant Dock.
>
> AL: That's your idea.
>
> LEM: Well, you don't think old man Walcott would sit around and let him get away with it, do you? For Christ sake, Al, put two and two together, will you?
>
> AL: Well, by god some nigger raped her.
>
> MARTY: Rape, my eye!
>
> AL: Yeah, rape. You let 'em get away with it, and no white woman will be safe on the streets any more. Christ, if a nigger raped your woman in your back yard, how would you feel if we helped him?
>
> LEM: Listen, Al, I don't stand for rape any more'n you do. But when anything smells as fishy as this does, I just know better, that's all.[47]

Echoing similar sentiments expressed by a black worker earlier in the play, Al says, "I never did savvy the idea of blacks and whites in one union anyway"—for which Lem also has a ready rejoinder, thus striking a decisive blow against "white chauvinism" among the rank-and-file:

> LEM: Listen, Al. We've had this out before. The only way we can tie up this river front is by organizing these black boys, and you know it. There are three of them to every white man on these docks. And if you think you're going to pull a strike in two weeks without them, you're crazy. Now, what do you want them to do? Stick with us, or scab for Walcott?

AL: They'll scab on you anyway. You can't trust those dirty niggers.

LEM: That's it! That's the stuff! You call them dirty niggers, and they call you low white trash. If you'd cut that out, if you'd get together and fight the Oceanic Stevedore Company for a change, maybe we'd get somewhere.

AL: Aw, you're talking through your hat.

LEM: Don't be such a god damned fool. You work for the Oceanic Stevedore Company just the same as they do. You get your wages cut just the same as they do. And every time they get it in the neck, you can be pretty damned sure you're going to get it in the neck too.

AL: You talk just like a nigger-lover.

LEM: Aw, for Christ's sake, Al.

AL: Yeah. Like a god damned nigger-lover.

LEM: Nigger-lover! We've been working day and night to build up this union. If we're ever going to get anywhere—if we're ever going to get a decent living wage, it's going to be through this union. And if fighting for this union makes me a nigger-lover, all right: that's what I am.[48]

Al storms off in anger, to reappear later in the play as a member of the white gang that terrorizes the black community, and the pace of the action rises. The black community rallies to conceal Lonnie; Lem Morris offers his assistance, but is greeted with skepticism and mistrust by some of the black community; Lonnie escapes two of his white pursuers and makes his way to Binnie's lunchroom, where the customers are listening to radio reports of white posses rampaging through black neighborhoods; Lem arrives at Binnie's, but is followed shortly after by a gang of white toughs; a melee breaks out; Lonnie escapes, but in the ensuing confusion, Sam, Ruby's father, is shot to death.

The final act of *Stevedore* is set in Bertha Williams's attic, where Sam's family and friends have gathered for his funeral service. Lonnie appears, followed shortly by Blacksnake, who informs the gathering that a white gang is preparing to attack the neighborhood. A spirited argument about what to do breaks out, with the preacher arguing vociferously for meekness and humility. This time, however, Lonnie prevails with a straightforward appeal to racial memory and racial solidarity:

Every time de white boss crack de whip you turn and run. You let him beat you, you let him hound you, you let him work you to death.

When you gwine to put a stop to it, black man? When you gwine
turn on 'em? When you gwine say: "You can't do dat. I'm a man. I
got de rights of a man. I'm gwine fight like a man."[49]

Lem arrives and explicitly addresses the widespread concern about whether
the black community will have to fight alone: "Well, Jesus, if you fellows are
going to fight that gang, you're going to need some help. . . . I'll go down and
call up Union Hall and see how many fellows I can round up. We'll give dat
Mitch boy a little more than he's looking for."[50] Buttressed by this promise of
support, the blacks erect barricades and prepare to defend themselves with
whatever weapons they have at their disposal: hot water, baseball bats, bricks,
chunks of coal, and several pistols and shotguns. Before the battle with the
white mob is joined, Lonnie rises to make his last exhortation:

> You all know what we hyar for. We hyar to defend our homes. We hyar
> to fight fo' our lives. And we hyar to show 'em dat we ain't gwine be
> kicked around, and starved and stepped on no mo'. We hyar to
> show 'em we men and gwine be treated like men. And remember
> we ain't only fighting fo' ourselves. Dar black folks all over de coun-
> try looking at us right now: dey counting on us, crying to us: "Stand
> yo' ground. You fighting fo' us. You fighting fo' all of us. . . ."[51]

Right after his speech, a shot rings out and Lonnie is killed—a sacrificial
martyr to the struggle he has organized but will not witness. For a moment it
appears that the black community will revert to its previous political paralysis—
"*A shriek of fear and despair goes up from the Negroes, who creep off toward the
corner of the courtyard*"—before Blacksnake leads them on. Shortly after,
Lem and a small band of union men join the fray; Binnie fires the pistol shot
that brings down the gang-leader Mitch; and the united force of black and
white workers routs the mob as Ruby, Lem, and a small group gather around
Lonnie's body.

According to many contemporary accounts, the last scene of *Stevedore* was
riveting theater. During a New York City performance, Bill "Bojangles" Robinson was reported to have jumped from his seat and mounted the stage to
cheer on the embattled black community.[52] A cast member who performed in
the Seattle Negro Unit of the Federal Theatre Project's 1936 production of
Stevedore later recalled the powerful reaction of an audience to the performance. As it happened, the play was staged during a longshoremen's strike
and many union members were in the audience. During the climactic scene,
some of them rushed the stage to help the cast erect the barricades against the
white mob.[53]

Several reviewers for the mainstream and liberal press poked gentle fun at
the ending of *Stevedore*, as did Joseph Wood Krutch, who observed in his *Nation* review: "Those brought up on melodrama may be reminded faintly of

the old days when the marines used to get there at the last minute."[54] But Krutch also confessed to being swept up in the action of the play, even as he acknowledged that its ending was more rooted in utopian hopes than historical reality, and praised the playwrights for sparing the black community from the white mob—"at least until after the curtain had fallen."[55]

The Communist press generally registered great satisfaction with *Stevedore*. Michael Gold was absolutely euphoric about the play,[56] but John Howard Lawson—himself a recent recruit to the Communist Party—wrote a retrospective complaint objecting to the absence of a clear-cut political line in John Wexley's *They Shall Not Die and Stevedore*, among other plays. In *Stevedore*, Lawson groused, Peters and Sklar had not sufficiently identified the union that had come to the aid of the black worker; some members might therefore conclude that the union was an AFL local, Lawson argued:

> It is absurd for any writer to attempt to write about the class struggle in *general* terms. The Communist Party is playing a definite role in every strike, in every activity of the working class. Other parties and groups are also playing definite roles. Every worker is intensely concerned with the policies of these parties. In fact, his life depends on a correct estimation of these policies.[57]

Speaking on behalf of the Theatre Union, Liston M. Oak rebutted Lawson's charges by reasserting the "united front" policy of the company. Although the character of Lem Morris was not called a Communist, Oak pointed out, he was called a "Red" and he was a "militant." Oak also chided Lawson for his concerns about the racial attitudes of the AFL, pointing out that some of its locals had risen to the assistance of African Americans in the past.[58] In spite of Lawson's quibbles, *Stevedore* was an unqualified success. After closing in New York it went on the road to Philadelphia, Washington, Detroit, and Chicago. It also played in London, with Paul Robeson starring as Lonnie Thompson.

Although *Stevedore* was technically not "about" the Scottsboro case, it effectively captured many of the social and political preoccupations of the plays that took Scottsboro as their subject—and in this regard can be seen as the high-water mark of works of this genre produced during the 1930s. For the Communist Party, and the Left in general, the fundamental challenge presented by Scottsboro was that of effectively bringing a Marxist analysis to bear on the plight of blacks in the United States, particularly in the South—how to dramatize the connection between the socioeconomic condition of blacks and capitalist exploitation.

First, in its explicit assertion that the false accusation of rape had to be understood in the context of the political and economic interests of the white ruling class, *Stevedore* sought to shift the grounds of the discourse away from the sexual titillation often associated with such accusations—recalling in the process some of the compelling arguments Ida B. Wells had raised in the late nineteenth century, but outside of the framework of Marxist analysis. By constructing the opening scene of the play in the manner they did, Peters and Sklar gave absolutely no credence to Florrie Reynolds's claims—and definitively established her as a character of ill-repute, thus clearing the space for a more thorough investigation of the *real* forces at work in their dramatic universe. In a sense, their efforts mirrored earlier debates among Communist Party strategists about how much attention should be given to respecting the sensibilities of southern white workers whose support they hoped to secure. For Peters and Sklar, the answer was clear, as we see through Lem's debate with Al: accept the facts, combat "white chauvinism."

Secondly, *Stevedore* carefully probed an issue that was central to the Scottsboro Narrative of the 1930s: under what conditions would African Americans rise up and strike out against the racial terror of southern Jim Crow? Langston Hughes, Muriel Rukeyser, John Wexley, and others had all raised this question too—often seizing upon the Scottsboro defendant Haywood Patterson, easily the most defiant of the Scottsboro Boys, as the figure in whom they invested their revolutionary hopes. Lonnie Thompson clearly embodies the qualities of revolutionary will that Peters and Sklar admire—even though he will finally die as a martyr. Although Lonnie's refusal to accept the rituals of the racial status quo are deeply grounded in his experiences as a southern black man, he is also the character who is most committed to forming biracial alliances to achieve social and economic justice.

Thirdly, *Stevedore* is committed to exploring the social and political logic behind the slogan "Black and White, Unite and Fight." Under what conditions can effective biracial alliances emerge? In *Stevedore*, Peters and Sklar are careful to indicate that, among the whites, the forces most hostile to blacks are the bosses, their representatives in the form of the legal authorities, and those workers who have been duped and misled by their own irrational and shortsighted racism. The white workers, led by Lem Morris, clearly are their friends. Similarly, within the black community there is also a wide spectrum of attitudes, ranging from the stock figure of the reactionary minister, to an old generation which—however reluctantly—has accepted the necessity of the racial status quo, to those who are deeply suspicious of whites. With few exceptions, however, these divisions are more fluid than those in the white community—as we see when the blacks close ranks during moments of social and political crisis. Peters and Sklar thus present an optimistic, indeed a triumphalist, vision of black-white solidarity at the conclusion of the play. Unlike the ending of Wexley's *They Shall Not Die*, the outcome of the play rests in the hands of black and white workers. The successful resolution of the Scottsboro case—the conclusion of *Stevedore* implies—would require similar efforts on the streets.

Some supporters of the Scottsboro Boys were quick to see the possibilities of the movie industry for shaping public opinion about the case. Clarence Muse, the noted African American actor who made his debut as the costar of the all-black musical *Hearts in Dixie* in 1929, urged Walter White to consider launching a film project to aid the Scottsboro defense:

> Now Walter here is what you are up against[:] the lack of interest in your organization at present and the I.L.D. are a little independent because of their success to date of arrousing [*sic*] the entire world in this Scottsboro matter, that is they take the credit.[59]

Muse seems particularly proud of the inroads he has made in the Jewish community:

> Using my same method I spoke to you about I immediately hopped on the Jewish program. I was a sensation on the largest affair ever given here by the Temple Israel, including artists like Will Rogers, Four Marx Bros., Bebe Daniels, Bing Crosby, Morton Downey and thirty others as famous. The result was that the next day at the studio, MGM they presented me with a brilliant idea to raise funds for our cause.[60]

In the wake of the acrimonious public split between the NAACP and the ILD, Muse proposes yet another coalition effort, focusing particularly upon the common interests of African Americans and Jews:

> If we can operate as a special committee, having the full support of the N.A.A.C.P., the National Urban League, the I.L.D., and associate organizations, Mr. Bonsky headed by several other MGM executives will make a picture, the profits to go to the cause.
>
> They have a clever story, with real heart interest, something we've never had in pictures concerning us, and it would be made right up to the standard with a mixed cast featuring me. If interested will send you a script.
>
> Here's the kick, no artists will receive any salary, our studio sets will be donated, director will serve as his bit, in other words with the Jews and Negroes of the country in on this, you could save your organization a lot of grief and at the same time maintain your high standard of serving the people; the boys' NECKS will be saved and the organization differences can be ironed out later.
>
> You understand Walter this is all in the rough, in fact I am feeling you and the Urban League out because the other gang will grab this idea in a minute, so it is up to you, Kinckle Jones, Carter, Hill to go into your chambers and think it out. In short, are you big enough to get together with the Jews who are also in trouble[,] the communists, the Church, in fact every agency to stop ALABAMA in her mad rush to the lynching trees. . . .[61]

By mid-May a prospectus had been drafted in cooperation with the Los Angeles Urban League, but the national office of the NAACP was slow to respond to Muse's appeal. A sense of the NAACP's attitude towards the project can be gleaned from a letter from Roy Wilkins to Claude Barnett of the Associated Negro Press in June 1933, complaining—among other things—about the lack of coordination among national organizations:

> The principal obstacle to our endorsement of the proposition is that it requires a promotion fund of $50,000. . . . There are other difficulties but this is the main one. No explanation has been made as to how the fund would be divided, if it were divided. The prospectus estimates that a half million dollars could be raised. Now this money would be raised on the strength of the Scottsboro case. The Scottsboro case will not cost that much money. What will be done with the balance? . . . Muse is very enthusiastic over the whole thing but he evidently thinks that money grows on trees, and is not conversant with the difficulties of raising this huge sum which is more than our annual budget.[62]

Walter White finally responded to Clarence Muse in mid-July, offering his standard litany of complaints about the ILD and declining to support the film project. White's letter to Muse crossed paths with a letter to White from Floyd Covington, Executive Secretary of the Los Angeles Urban League and an enthusiastic supporter of Muse's project, appealing for an answer. When White repeated to Covington what he had already written to Muse, Covington graciously accepted his refusal, but concluded his letter on a poignant note, mourning a significant opportunity for African Americans and Jews to collaborate on a project of mutual concern:

> Because of the situation in Germany the Jewish people are very anxious to combine with all minor groups that are undergoing a similar condition. If a national and open defense fund should be established, it might enable the Negro, along with the Jewish people, to produce pictures commercially, which would give him the money he might need in his own defense.[63]

Even if the irreparable breach between the ILD and the NAACP had not existed, the prospect of producing a film about the Scottsboro case, with a multiracial cast and prominently featuring Clarence Muse, would have been daunting in 1930s Hollywood.

When the Scottsboro case intruded upon mainstream cinema, it did so indirectly and covertly—as in the case of director William A. "Wild Man" Wellman's 1933 production *Wild Boys of the Road*. Bracketed between his groundbreaking 1927 film *Wings*, which earned the distinction of winning the first Oscar for Best Picture, and his somber 1943 production, *The Ox-Bow Incident*, also nominated for Best Picture, Wellman added to the roster of socially conscious films produced by Warner Brothers Studios with this effort.

Wild Boys of the Road begins rather innocuously, with a jazz age overture that revolves around youthful high jinks of two of the main characters, Eddie and Tommy, at a high school dance in a small town somewhere west of the Mississippi. Even in the midst of their pranks, however, there are signs of economic troubles: Tommy lacks the funds to pay for admission to the dance and is reluctant to go downtown. He confesses that his mother has been relying upon the Community Chest for their meals and is now working downtown to make ends meet. Eddie seems better off financially, but he returns home late that evening to discover his parents fretting over their future; his father, they announce, has just been laid off. Several months pass and Tommy and Eddie decide to drop out of high school and take to the road to lessen the economic burden on their families. From the point that they hop on a freight train passing through their town, *Wild Boys of the Road* takes on the gritty texture of other 1930s social problem films such as King Vidor's *Our Daily Bread*.

Early in their journey on the freight train, Eddie and Tommy meet a young woman, Sally, traveling solo and disguised in men's clothing. "He's a she!" Eddie exclaims after making the discovery. The three quickly become fast friends as they travel by freight train across the country, joining with many other young people—most of them white, but some blacks as well—often eluding the clutches of the railroad police and living by their wits. As the film wends its way across the country, heading east, it graphically maps a landscape ravaged by the stock market collapse and the Great Depression. In a critical scene in the film, hundreds of teenagers express their solidarity by launching a ferocious counterattack against the railroad police who have routinely harassed them. Meanwhile, a young woman who has remained on a freight car is raped by a white railroad worker. In rapid succession the army of young people beat up her assailant and throw him from the train, perhaps killing him, and Tommy loses his leg when he falls in front of an oncoming train. After their expulsion from an edenic "sewer-pipe city" outside of Cleveland, Ohio, the band of young people disperses and Eddie, Tommy, and Sally make their way to New York City, where further adventures await—culminating in Eddie's arrest on a range of charges. Earlier scenes in *Wild Boys of the Road* made a point of underscoring that some of the stern authority figures encountered by the young people on the road regarded them with some degree of sympathy, and the final scenes of the film introduce yet another sympathetic figure in the person of the judge who tries Eddie. Eddie, who has been distinguished by his toughness throughout the film, stands up to the judge with an impassioned speech: "I'll tell you why we can't go home: because our folks are poor, they can't get jobs, and there isn't enough. . . . You're sending us to jail because you don't want to see us; you want to forget us. Well, you can't do it. Because I'm not the only one; there's thousands like me—and there's more hitting the road every day." With the NRA Eagle, the symbol of Franklin D. Roosevelt's first administration's efforts to combat the

Depression, squarely on the wall behind him, the judge gently responds to Eddie's tirade: "I'm going to dismiss your case. . . . Furthermore, I'm going to help you." And, echoing the slogan of the NRA ("We Do Our Part"), the judge tells Eddie, Tommy, and Sally: "I'm going to do my part and now I want you to do yours. . . . Things are going to be better now—not only here in New York, but all over the country." And so *Wild Boys of the Road* ends on a triumphal note, celebrating the promise of the New Deal.

"F.S.N.," a reviewer for the *New York Times*, was deeply disappointed by the film, arguing that it "might have been at once tragic, dramatic, and a stirring call for remedial action. Somehow it has missed being any of those. Its tragedy has been oversentimentalized, its drama is mostly melodrama and, by endowing it with a happy ending, the producers have robbed it of its value as a social challenge." Wellman, "F.S.N." concluded, "has taken a theme with broad social implications and has converted it into a rather pointless yarn about three wandering youngsters."[64]

What "F.S.N." overlooked, however, and what some discerning 1930s movie viewers may have noted, are the ways in which *Wild Boys of the Road* seemed to indirectly comment on the Scottsboro case and the fate of the Scottsboro Boys. In effect, Wellman took the basic elements of the Scottsboro Narrative—the dire economic conditions that propelled many young Americans, men and women, on freight trains across America in search of better economic opportunities; the presence of racially mixed communities on these journeys; the report of rape in a racially charged context; vigilante justice; and the intervention of legal authorities—and dramatically refashioned them. *Wild Boys of the Road* is particularly noteworthy in its incorporation of black characters in this regard. Although young African American males constitute a distinct minority among the growing band of youth who are riding the rails, their presence is not simply that of fulfilling the comic function usually conferred upon African Americans by 1930s Hollywood conventions (even though they *do* perform this function in a scene in which they smash watermelons on the heads of pursuing railroad authorities); rather, they receive essentially the same treatment by the white, adult figures in the film as everyone else. Their presence is assumed, in other words, as one component of a shared economic existence. Moreover, the rape scene—which occurs off-camera—makes it plain that the villain (played by the then unknown actor, Ward Bond) is a white man. In the rough vigilante justice that immediately follows, the black youth participate as well. Characters and situations that could have emerged as sources of dramatic tension in *Wild Boys of the Road* are displaced by Wellman's focus upon teenagers caught up in the grips of the Depression.

In short, some of the same social and historical conditions—with the critical exception of the geographical setting—that created the Scottsboro case provided Wellman the opportunity to imagine another outcome, not only for his characters, but for Americans under the benign guidance of FDR's New Deal.

On the surface at least, *Fury*, the first American film directed by the Austrian-born émigré Fritz Lang, seemed to offer more direct commentary on the Scottsboro case. Lang had fled from Nazi Germany in 1933, arriving in Hollywood the following year. After two years under contract to MGM Studios, during which he struggled to adapt to the English language, American culture, and the Hollywood studio system, Lang produced the film still regarded by many critics as the finest film he directed in the United States. *Fury* was based upon an incident of mob violence that had occurred in San Jose, California on November 27, 1933:

> a mob of vigilantes had broken into a jail . . . and lynched two suspects in the kidnapping and murder of a widely known department-store heir. News bulletins and photos of the hangings—the dangling victims stripped nude—stunned the nation. There were several unique angles: The kidnapping was never satisfactorily solved, with the alleged ringleader, one of the two lynching victims, maintaining his innocence up to the bitter end. The governor of California shamed himself by endorsing the lynching. And when the lynching case was brought to court, the locals stuck together, refusing to identify the mob members. Only one person among hundreds was indicted, the charges dismissed in February of 1934.[65]

The script, initially called "Mob Rule," went through several transformations before it was finally approved for filming in April and May of 1936. *Fury* begins with a series of Depression-era vignettes. Joe Wilson (Spencer Tracy) and Katherine Grant (Sylvia Sidney) are presented as decent, honest, hard-working people who are in love but do not have enough money to get married. Katherine leaves their midwestern hometown to take a job in another city, while Joe remains behind, hoping to save enough money to buy a gas station to support him and Katherine after they are married. A year later, Joe is stopped by a deputy sheriff in a small town as he drives across the country to meet Katherine for their impending wedding. A kidnapping has occurred in the town and circumstantial evidence apparently links Joe to the crime. Arrested and jailed in spite of his vehement protests, Joe sits in his jail cell while rumors spread throughout the town. A lynch mob forms and storms the jail. Unable to breach the walls of the building, the mob sets the jail on fire. Joe is presumed to have died in the conflagration, but somehow he manages to escape. Impelled by his fury at the mob ("I could smell myself burning," he later recounts), Joe spends the rest of the film trying to exact revenge upon his tormentors, working behind the scenes and enlisting the support of his brothers, whom he has drawn into his scheme. Twenty-two of the vigilantes have subsequently been arrested for their actions and Joe is determined to see them sentenced for his murder. Katherine discovers that he is still alive, however, and confronts him, appealing to his conscience and sense of decency. In a dramatic, last-minute appearance, Joe reveals himself to the courtroom, lashing out at the blood-lust, cowardice, and inhumanity of the townspeople, but not apologizing for his own behavior; at the end, he is a bitter and spiritually broken man.

The broad analogies between *Fury* and the Scottsboro case seemed clear to some viewers. An attorney, Arthur Hartley, wrote to William Jay Schieffelin of the American Scottsboro Committee immediately after seeing the movie in June 1936:

> As you have undoubtedly heard, this motion picture shows from beginning to end how a lynch spirit is whipped up, and an innocent man is victimized. Its realism is startling and it can safely be said that it is one of the greatest movies produced by Hollywood touching on social questions.
>
> I am writing you to suggest that the showing of this movie should be utilized by the Scottsboro committee for the furtherance of its work. I think that a leaflet or a pamphlet should be gotten out at once tying up the Scottsboro case with the lessons of the movie. I think that through the United Citizens Committee of the American League against War and Fascism, of which I am a member, some contact might be made with the theatre permitting the distribution in its lobby.[66]

Lang himself, the author of so many of the myths that continue to circulate about him, would subsequently claim that in an earlier stage of the script "he had tried to turn *Fury* into a film about a black man accused of raping a white woman"[67]—an assertion that recent biographer Patrick Gilligan dismisses out of hand as "especially absurd, given both the historical record and the social environment."[68] Gilligan acknowledges that the experience of making *Fury* would have certainly had the effect of educating Lang about America's racial history, but

> It was always Spencer Tracy, never a black actor; nor could it have been a black leading man at MGM or any other major Hollywood studio in 1935, when the film studios were still deeply beholden to the theatre-owners and segregationist exhibitors in the Deep South.[69]

The more time passed the more deeply convinced Lang became, apparently, that he had indeed intended to direct a film about a falsely accused African American victim—only to be thwarted by the machinations of a socially and politically reactionary studio system. *Fury* conveyed its sharp indictment of mob rule with striking visual power. Still, it would take several more decades before Hollywood was prepared to place African American subjects directly at the center of the kind of script that Fritz Lang imagined.

||||||||||||||||||||||||||||||||||||||||||||||||||||||||||||||||||||||||||||||||||||||||||||||||||||||||||||||||||

CHAPTER FOUR

FICTIONAL SCOTTSBOROS

NOVELISTS WHO SOUGHT to tackle the subject of the Scottsboro case faced more formidable artistic and political challenges than their counterparts who wrote poetry or drama, particularly since the gap between historical events and fictional composition could be difficult to navigate. In the mid-1930s the outcome of the Scottsboro case was far from resolved; nor could the defenders of the Scottsboro Boys predict their future with any degree of confidence. Novelists who subscribed to the norms of realistic fiction were therefore somewhat circumscribed by the actual circumstances of the case. Nevertheless, three writers turned to fiction to convey their involvement with the Scottsboro case: Grace Lumpkin, Guy Endore, and Arna Bontemps.

Born in 1891 in Milledgeville, Georgia, the sixth of seven children, Grace Lumpkin belonged to a once-prominent Georgia family whose fortunes had sharply declined since the end of the Civil War. Her father, William Wallace Lumpkin, who invested considerable emotional and intellectual capital in the memory of the Confederacy and the "Lost Cause," moved the family to South Carolina, where Lumpkin (along with her younger sister, Katharine Du Pre, later to become famous for her 1946 memoir *The Making of a Southerner*) grew up as an active participant in the public rituals of the United Daughters of the Confederacy. She was educated at Brenau College in Georgia, and began working at a variety of jobs that helped to shape her nascent political consciousness: teaching school; organizing night school for farmers and their wives; working with textile workers in the mountains of North Carolina; and working for the YWCA, where she became convinced that "workers could only better their lives by means of unions."[1] Through her work with the YWCA Lumpkin was exposed to the most progressive opinion about race relations available to women of her race, class, and generation. After her mother's death in 1925, she used her YWCA connections to land a job in New York City as a member of the staff of *The World Tomorrow*, a monthly Quaker publication—a way station on her road towards the Left. Lumpkin covered the long, often violent textile workers' strike in Passaic, New Jersey for *The World Tomorrow* and became more and more attracted to Communist Party politics. Although she never joined the Party, she became a dedicated fellow-traveler. The Greenwich Village apartment she shared with her roommate Esther Shemitz soon became a gathering place for a small group of Communists, including Michael Intrator, who subsequently became Lumpkin's lover, and Whittaker Chambers, then an active member of the Communist Party, later the notorious informer in the celebrated Alger Hiss trial of the late 1940s.

Lumpkin published her first sketch, "White Man—A Story," in *The New Masses*; in 1927 she was arrested on a picket line protesting the Sacco and Vanzetti verdicts; in 1928, she joined the staff of *The New Masses*; in 1929, she went South to observe and participate in the Gastonia, North Carolina textile workers' strikes—experiences that she subsequently incorporated into her first novel *To Make My Bread*, the recipient of the 1932 Maxim Gorky Award for the best labor novel. Lumpkin was also deeply engaged with the Scottsboro case, publishing a number of firsthand accounts in the International Labor Defense's *Labor Defender*. She published her second novel, *A Sign for Cain*, in 1935.

Set in a small southern town in the early 1930s, *A Sign for Cain* is in many respects a melodramatic potboiler, complete with a large cast of stock characters and settings from postbellum southern fiction. The novel revolves around the Gault family, once members of the town's social elite but now experiencing the final spiral of an economic decline that began at the end of the Civil War. The Gaults are presided over by Colonel Gault, now dying, whose manners and outlook have been decisively shaped by the legacy of slavery, the Civil War, and the "Lost Cause." The Colonel is deeply estranged from his sister Evelyn—who had the temerity to marry Judson Gardner, a political rival and the son of a poor white family whose father became wealthy selling goods to "carpet-baggers and scalawags and nigras." The Colonel has never forgotten his ignominious defeat by Gardner in their race for the governorship many years before and has never forgiven his sister for marrying him. Colonel Gault continues to exert his patriarchal authority over his son Charles, the minister of a church in town, and his wife, Louise; his son Jim, a dissolute ne'er-do-well who has an illicit relationship with a young black woman in nearby Junction City; and—later in the novel—his daughter Caroline, a successful novelist who has returned home from New York to care for her father.

The plantation setting of the Gault household includes an old slave cabin where Nancy, an old servant for the Gaults and a model of Christian forbearance, lives. Like Colonel Gault, she is ailing, and the novel begins with her dictating a letter to her son Denis, who left home for the North many years ago. Her immediate community includes her minister, Brother Morton; and the remaining black sharecroppers on the Gault land: her young neighbor Ficents, and a friend of her late husband, Ed Clarke—both of whom will have important roles to play in the unfolding plot of the novel.

Other key figures in *A Sign for Cain* include Bill Duncan, a distant cousin of the Gaults and the editor of the *Jefferson Record*, the newspaper he has inherited from his father; Selah, a young black woman, Ficents's girlfriend, who has been "bound out" to the Browdie rooming house; Lee Foster, a white farmer who has lost his farm and is now renting the overseer's house on Gault's land; Ross Sellers, a disreputable white who lives near Ficents and makes his living selling illegal liquor, and his daughter, Mary; Jud Gardner, the son of Evelyn's late husband by his first wife; Walt Anderson, the typesetter for the *Jefferson Record* and a long-time committed Socialist; Muriel Browdie, the social reporter for the *Jefferson Record* who has recently returned from New York City; and the prosperous and influential Judge Bell.

Lumpkin's anatomy of this community quickly reveals a world sharply divided along class, racial, and ideological lines. When Denis returns to take care of his ailing mother Nancy, he has an apparently chance encounter with Bill Duncan, the newspaper editor, that leads to the following exchange:

> "What do you think about it?" Bill asked. "I mean do you think things will get better?
>
> "They might for a while, but the hard times are bound to come back. What do you think?"
>
> "About the same. It seems as if the poor will get more so, and the rich more so and the middle ones will lose most of what they have."
>
> The young Negro gave a quick sideways glance at Bill. "You been studying about that, too?" he asked.
>
> "I was thinking about that before you two were born," Walt told them.
>
> "Walt's a Socialist," Bill explained. He looked at Denis as if he expected some special recognition of the meaning of what he had said. But Denis's expression did not change.[2]

Lumpkin thus establishes Denis and Bill's credentials as Communists with a shared—and dangerous—mission in the region. (Significantly, and consistent with the Communist Party's "Third Period" politics, Walt—the Socialist—does not play a significant role in the decisive action of the novel.) And the mission is to organize sharecroppers and other workers in the region—in spite of Nancy's fervent appeal to Denis that he accept his place in the established racial order as a natural feature of their universe:

> "We got to take what the Lord sends and be thankful. What you trying to do, Denis?" she asked querulously.
>
> "We don't have to take what the Lord sends. I wonder . . . even if . . ."
>
> "I knows what I'm talking about," Nancy's voice was stronger. "It ain't no use, Denis. The only way to help yourself is to get white folks to help you. If you can't be a preacher the Colonel will help you get what you want,

maybe a little farm or maybe set you up with a pressing club in town. You got to depend on the white folks. I thrashed that out long ago. . . ."

Presently he said, without looking at Nancy, "I told you I got to work for my people."

"And I told you you could do that preaching to them."

"It's not what I mean. My people are white people, too. They are working people, white and colored."

"You mean po' white people, white trash is *your* people?"

Denis nodded.

"White trash is your people?" she repeated.

"I mean white working people."

"In all my days I never heered of such a thing . . . in all my days," Nancy whispered.[3]

The network of Communists and their supporters soon extends to Lee Foster, Ficents, Selah, Ed Clarke, and others; and the scope of their organizing efforts extends from sharecroppers, black and white, to black women launderers in Gaulttown, the black section of town. These efforts, and the visible shifts in social relationships between blacks and whites as a result of them, constitute the decisive backdrop for the major action of the novel. As Sheriff Harrison makes clear in a visit to Colonel Gault, the work of Denis and others seriously threatens deeply entrenched social and economic interests:

> "I heard in a round-a-bout way . . . that your nigger Denis is making trouble around here with the niggers. Now, Colonel, I've made this here community a peaceful one. We don't want trouble like they've had in other states. I've kept the peace. You know that. Ain't been a nigger around here for years but has been scared to raise his voice too high. . . . And I don't want no trouble," his voice became a whine, "and some say the whites are getting it too."[4]

Moreover, the Sheriff adds, Denis has effectively mobilized the black launderers, too:

> "Henry Bandy, he owns this big laundry you know, and he brought up in town council that people shouldn't have nigger wash women. He thought it was a bad thing for the town because his laundry is sanitary and everybody knows you get disease and dirt and . . . insects from the nigger wash women.
>
> "So he got me to send a health officer over to Gaulttown to inspect and the health officer said it wasn't safe for people to send their washing over there. Well, we heard the niggers was complaining about it. And they had a meeting in the church. The nigger preacher Morton came to us about that. He said he didn't know beforehand what the meeting was to be about. He's a good nigger and was afraid we would blame it on him."[5]

Sheriff Harrison not only exposes the complicity of the preacher Morton in the oppression of the black community, he also reveals the extent to which the *Communist Manifesto* and other documents are now circulating within it. As a result, Denis—like Lonnie Thompson in Peters and Sklar's *Stevedore*—is officially designated as a disturber of the peace, a Communist organizer whose work threatens the existing race rituals of the town.

Evelyn Gardner is murdered by her nephew, Jim Gault, midway through *A Sign for Cain*, and, from this point on, Lumpkin plots her novel around deliberate parallels with the Scottsboro case. In a moment of intense self-revelation, Lee Foster, who has discovered Evelyn's body, confesses to Bill Duncan his own conflicting racial attitudes:

> "I told Denis they better lie low," Lee said to Bill when he had finished with the story. "You know all the niggers on that place will be suspected. It's the natural thing, not natural, but usual, I reckon.
>
> "I, myself, I . . . I . . ." he looked shamefacedly at Bill. "There she was," he said, "lying on the ground with two holes in her forehead and the hair matted, and the rock lying near by covered with blood. And right away it came to me, 'a nigger did it,' and a rage came up in me against them, all of them. That's what we been taught, Bill, the same as you. It's in the air—we breathe it like air. . . . Just like that when something like this happens you say 'rape' and run to find the nigger.
>
> "I got up from leaning over her and started out full of rage. Then I remembered something you told me, 'Maybe there are some like that, but if there are it's the fault of the people who keep them living like beasts. We got to be men and think like men and not be unthinking animals.' So I turned around and looked again. I looked with reason in my mind."[6]

As in the Scottsboro case, word of Evelyn Gardner's murder circulates in town on a trading day, when a crowd of country people are milling around. And Sheriff Harrison is quick to insinuate that an act of rape has been committed. On the surface, the country people who begin to assemble expectantly around the county courthouse seem to bear a sharp resemblance to the townspeople in Rukeyser's "The Lynchings of Jesus" and Wexley's *They Shall Not Die*, but Lumpkin tends to view them more sympathetically; they are not hopelessly unredeemable:

> Just below, on the next granite step sat a long lean farmer with cheek bones jutting out and deep hollows under them. He was in overalls patched at the knees. His face was sallow and unhealthy looking. . . .
>
> Most of the people on the court house lawn were rather poorly dressed, and on many of the faces there was an expression of blank hopelessness. Yet Bill, who had talked with some of them, knew that in many cases this expression was deceptive. These were people whose ancestors had cleared

out the forests and made homes in a wilderness. They had cleared the wilderness and others had come and taken it from them. Gradually, through the generations all their hopes for themselves and their families had become frustrated.

The frustrated hopes had produced a hidden anger, which could be turned in any direction, into any channel. . . .

Bill looked at the farmer and thought how easily the hidden unknown anger caused by frustrated hopes could be directed toward people of another race.[7]

In sharp contrast, the sheriff's deputies are depicted as genetically defective throwbacks:

Except for the sheriff and two others they were faces of men in their late twenties or early thirties, yet they appeared bloated like the faces of diseased old men and the harsh mouth lines only emphasized the effeminate, self-indulgent expression of their lips. Even Boyle, the tall deputy at the head of the table, in spite of his muscular body had a weak, unmanly chin.[8]

Although Bill does his best to thwart the quickly emerging public consensus about a rape-murder, he is unsuccessful in challenging the story the sheriff has concocted. Mary Sellers—the poor white equivalent of the actual Ruby Bates and the counterpart of Wexley's Lucy Price in *They Shall Not Die*—is carefully guided and coached about the story she should tell by Sheriff Harrison, who applies comparable inducements of intimidation and enhanced social status in exchange for her testimony. And word of her accusations considerably improves her standing in the white community:

Mr. Browdie was saying, "That poor little Mary Sellers. She must have been scared to death, finding them niggers doing that dastardly thing."

"I heard they come after her, too, and she fought them off and run," a man said. . . .

"I don't doubt it. I hear the sheriff says she's a grand little woman."

"I always did think she was more sinned against than sinning," another man said unctuously.

"Poor little girl, she must have gone through a lot. Them damned niggers. Burnin' is too good for them."[9]

Denis and Ficents are arrested for the crime and Bill telegrams his comrades in the North with news of the legal travesty. Bill's actions are absolutely consistent with the two-pronged strategy of the International Labor Defense in relationship to the Scottsboro case—including his anticipation of the legal challenge that led to the 1935 Supreme Court decision in *Norris v. Alabama*:

His common sense told him that until the lawyers came nothing definite could be done, except see that the world heard of what was happening, and watch carefully so that nothing would happen to the prisoners. When the lawyers arrived he and his group of farmers could help in getting witnesses and in the fight to put Negroes on the jury. As a child he had accepted the fact that Negroes were not allowed to take their constitutional right of being judged by members of their own race. As a Communist he had learned that the law of the country was being broken every day by his countrymen.[10]

In the meantime, Denis and Ficents—like the Scottsboro Boys—are savagely beaten by the legal authorities, and Ficents signs a confession under extreme duress. Telegrams holding Judge Bell responsible for the treatment of the two prisoners begin to pour into the Judge's residence, and his outrage casts serious doubt upon his sense of impartiality:

> "The God damn bastards," he shouted, "the God damn sons of bitches." His face was like red flannel. All the simplicity and gentleness which was characteristic of him was gone. "I'd like to kill them," he shouted. Raising his hand with the telegram high in the air, he shook it at Bill.[11]

To underscore the parallels with the Scottsboro case, Lumpkin stages a struggle between Bill Duncan and his now estranged lover, Caroline Gault, over which lawyers can best represent Denis and Ficents—a scene that replicates, on a smaller scale, the confusion, fear, and uncertainty of the families of the Scottsboro defendants:

> Bill turned to Ficents' mother. "Before the court will allow us to have these lawyers, you and Mum Nancy must say you want them," he explained. "It means that you give your consent to having our lawyers take the case. Will you do it?"
>
> "I certainly will. I certainly do want your lawyers to take the case," Rosa Williams said.
>
> "They are your lawyers," Bill said, "yours and Nancy's. . . ."
>
> "Mum Nancy," he said patiently, "as you know, Denis is in jail. . . ."
>
> "Oh, Bill," Caroline said reproachfully.
>
> "I know, but she must realize how everything depends on this. It is our only hope. Our only hope of getting him free," he said to Nancy, "is to get these lawyers.
>
> "Judge Bell said he would appoint lawyers, Bill," Caroline said in a clear voice. "He will get the best ones possible."
>
> "But don't you understand what that means, Caroline? Those lawyers appointed by the Judge won't really try to free them. I've seen too many cases of this kind. And you know them, too. We must get these others," his

words dried in his throat, for he had seen an unmistakable look of hatred pass over Caroline's face. . . .

"It is to save Denis," Bill reminded Nancy. "To save his life," he said gently, distinctly. "Don't you trust me, Mum Nancy?"

"I do trust you," she said, "I helped to raise you. But I don't know. I got to have the Colonel's word. I can't do nothing. If Miss Caroline says all right, then . . . Oh, My Lord, if they only hadn't done it . . ."[12]

In spite of the determined efforts of Bill and his embattled white allies, nothing can avert the implacable motion of the novel: goaded by several of his contemporaries, a drunken Jim Gault—clearly with the complicity of the deputies—shoots Denis and Ficents to death. In the final scenes of A Sign for Cain, Walt, the old Socialist, undergoes a political conversion:

"But I want to be at that meeting tonight," Walt said.

"I didn't know you'd want to come."

"I do."

"We are to have a public funeral, you know, Negroes and whites," Bill said, "and we intend to march down the streets. We're getting people from Gaulttown and Gardners. It may mean trouble."

"I know it."

"You mean this?" Bill asked.

"Yes. It's time to fight. I know it."

"I'm glad of that," Bill said. He rose from his chair, went over to Walt and held out his hand.[13]

Meanwhile, blacks and whites gather at Ed Clarke's cabin, planning their future actions. The scene has a martial air to it, and the quiet determination of the gathering points to a promising future—even if it is not depicted in the apocalyptic terms of the endings of Langston Hughes's Scottsboro Limited or Paul Peters and George Sklar's Stevedore:

The feet walked firmly. The light from the lantern spread over the men as they passed through the back door. It made them appear larger, of heroic proportions, and their shadows thrown against the white-washed wall of the cabin made it seem that others were walking with them, so that their number was multiplied.[14]

Lumpkin juxtaposes this scene with a comparable gathering at the Gault plantation, where the Colonel in his delirium shouts out snatches from his former speeches as his family spirals towards its final decline.

A Sign for Cain was published to decidedly mixed reviews, with the Boston Transcript praising her efforts:

Grace Lumpkin has shown that she can write entertaining radical propaganda. Entertaining because the propaganda is contained in the action rather than in windy diatribes by either the author or her characters. Her convictions about the class struggle are deep and sincere, but because of her wisdom or her unsubmergible ability to write, her work is primarily valuable for its readability.[15]

Erskine Caldwell's praise was more studied and qualified:

Grace Lumpkin has written a fore-runner of the revolutionary novel of the near future . . . After having admired her purpose, her sincerity, and her accomplishment, one cannot keep from suggesting what appear to be possible pitfalls. One of these possible, but not necessarily probable, pitfalls is the danger of falling into the employment of cant and repetitious formula. . . . Another pitfall is in the tendency of a propaganda author to stack the cards against the villains so hopelessly that the outcome of the conflict is never in doubt after page three.[16]

Writing for the *New Republic*, Robert Cantwell, whose novel *The Land of Plenty* has been acclaimed by some critics as one of the finest novels produced during the Depression, was less generous:

Its weaknesses as a novel are obvious, and arise principally from the fact that the characters are seen too narrowly in their social roles, that they are too consistently true to type, without the elements of waywardness and unpredictability that are as truly human possessions as eyes or class backgrounds. "A Sign for Cain" is in fact a kind of skeleton of a major work, the characters firmly outlined and located, a central situation carefully built up; but it remains a skeleton, for the living flesh of unexpected insights and poetic intuitions is lacking.[17]

The *New York Times* simply dismissed Lumpkin's efforts out of hand: "A Sign for Cain" is a novel that, once read, fades very rapidly in the mind. One retains from it a sense of its honesty, of its fervent condition . . . but the story and the people, in their relations to one another, blur.[18]

And fade it did. There is no evidence that *A Sign for Cain* had any discernible impact on public discourse about the Scottsboro case. In her later life, after she became a rabid anti-communist, Lumpkin placed considerable distance between herself and the novel, claiming that it had been written under the threat of Party discipline if she strayed from the Party line.[19]

While Grace Lumpkin recast the Scottsboro case into contemporary melodrama, both Guy Endore's *Babouk* (1934) and Arna Bontemps's *Black Thunder* (1936) reconstituted it as historical fiction.

Born Samuel Goldstein in New York City, Endore was raised under rather bizarre circumstances. According to Alan Wald:

His father, Isidor Goldstein, had been a coal miner in Pittsburgh, and his mother, Malka Halpern Goldstein, committed suicide four years after her son's birth, partly in response to the family's dire poverty. Soon after, Endore's father changed the family name to obliterate the past, placing the children in a Methodist orphanage. But then his father sold an invention and dreamed that his dead wife commanded him to use the money to obtain a European education for their children. This inaugurated a bizarre five-year sojourn in Vienna where the Endore children were left under the care of a Catholic governess. When the elder Endore vanished and the money ran out, Guy and his siblings were sent back to Pittsburgh to live communally.[20]

Endore somehow managed to attend Columbia University, where he majored in European languages, earning a B.A. in 1923 and an M.A. in 1925. While at Columbia he fell under the spell of the apparently ubiquitous Whittaker Chambers—the same figure who was closely associated with Grace Lumpkin during the 1920s and 1930s—who helped to prod him towards the Left. By the early 1930s, Alan Wald reports, "Endore had read everything by Marx and Engels available in English, French, and German, as well as Werner Sombart's multivolume history of capitalism."[21] When he published *Babouk* in 1934, hard on the heels of his best-selling novel about the Paris Commune, *The Werewolf of Paris* (1933), he was thirty-four years old and already the author of four other books and several translations. He had not yet joined the Communist Party—although he would do so in Hollywood between 1936 and 1938—and his passionate pamphlet in defense of the Scottsboro Boys, *The Crime at Scottsboro*,[22] had not yet been written. Endore's long and successful career as a novelist, Hollywood screen writer, and committed Communist was still ahead of him, but in *Babouk* Endore left no doubt about where his political sympathies rested.

It is ironic that *Babouk* was published the same year that the United States ended its nineteen-year military occupation of Haiti, and four years before the publication of C.L.R. James's justly praised classic account of the Haitian Revolution *The Black Jacobins*. The novel—on the surface at least—represents a bold and imaginative attempt both to challenge the historiographical silences about the brutal realities of the transatlantic slave trade and to recuperate the buried tradition of spirited African resistance to it in the French colony of Saint-Domingue, present-day Haiti.

From the opening graphic portrayal of the "nigger-taster"—whose distinctly unhygienic task was to lick each African slave and taste his sweat to determine his health and potential value—to the final dire drumbeat of warnings about impending black revolution to unthinking whites, *Babouk* offers an unprecedented account of the eighteenth-century slavery as it was experienced by the Africans themselves.

Babouk is based loosely upon the figure of the African priest and leader Boukman, whose 1791 voodoo ceremony is believed to have been the beginning of the slave resurrection that led to the Haitian Revolution. Little is known about the pre-revolutionary life of the historic Boukman, giving Endore virtual free rein for his imagination and considerable artistic talent.

Most commentators have noted that *Babouk* is distinguished by its multiple-voiced narration.[23] One narrative voice is omniscient, firmly grounded in a meticulous command of historical facts and documents. This voice tells the story of Babouk, a Mandingo and a born storyteller and philosopher, who is captured on the west coast of Africa when he is barely sixteen years old; it traces his journey from the harrowing Middle Passage through his sale and enslavement in eighteenth-century Saint-Domingue; his slow, torturous accommodation to the demands of slavery in the colony; his years of silent suffering until his emergence as a mature and respected storyteller and leader; and his apotheosis as a rebel leader whose efforts pave the way for the Haitian Revolution. Along the way, this narrative voice offers a detailed anatomy of plantation society—positioning readers to experience the stench and horrors of the Middle Passage, offering detailed descriptions of the mundane business details of the traffic in human bodies and glimpses of the behavior and conversations of the planter class. Endore's broad and detailed command of the history of African peoples and the transatlantic slave trade is readily apparent in his deft handling of ethnographical details, particularly his understanding of the ways in which the slave traders exploit ethnic and cultural differences to maintain and reinforce divisions within the slave community. Scathing in its portrayal of white slave traders and planters, this narrative voice is also unsentimental in its depiction of the broad range of humanity—with all of its frailties and imperfections—represented by the Africans.

The other narrative voice is firmly grounded in the world of the contemporary 1930s. Alternately wry, sardonic, ironic, indignant, and angry, this voice insists upon making bold connections between the historic commentary and the modern world, and takes delight in drawing attention to such anachronistic interventions. In one of its first sustained meditations, this narrator offers the following commentary about the public execution of two African men and one woman "convicted of having employed poison in the slaying of animals and slaves, and even of one white." Noting the sharp discrepancy between the way this public spectacle has been understood by blacks—many of whom see it as proof of the white man's cannibalism—and whites, he caustically observes:

> Fortunate is the student of history who, with documents in hand, can declare: Here in these three Negroes burnt at the stake is proof of the white man's indulgence.
>
> Yes, the white man's indulgence.

Stop being sentimental. Don't let pity blind you. Your heart must not be your judge.

Study history objectively, from documents.

But you can't study the actual legal documents on the case for the official papers of *Saint-Domingue* were largely destroyed every few years. Time and time again we find an entry in the court records: "Burnt the papers relating to Negroes, along with other useless documents. . . ."

But, the narrator continues, we can find fleeting references to Negroes in the volumes of letters written by one Ordinator Lambert, who offers the following comment about a similar execution: "We have reasons for believing that the cause of all this is too many liberties allowed the Negroes."[24]

There are other plausible hypotheses, the narrator indicates, other motives that have nothing to do with the "white man's indulgence" and everything to do with the fierce and abiding hope for freedom among the Africans:

The indulgent white master promises a Negress who has borne him a child that he will free her in his will.

The informed Negress at once schemes to poison her kind master and thus hasten his demise and her freedom. . . .

Another indulgent master, an absentee owner, is going to Paris and will leave an overseer in charge, that is to say, a man who will strive to grind out a little profit for himself. Quickly a wise witch-doctor on the plantation poisons a hundred slaves, or a hundred head of cattle, ruins the good master and prevents his departure.[25]

Then again, there may be other ways of understanding the "historical record":

Of course there's the possibility that the white master did not die of poison. That the slaves or the cattle died of some obscure epidemic. But that's another matter and hardly worth going into, seeing that it would lead us astray into the whole subject of whether it's right or wrong to punish a Negro for a crime he did not commit, for example, to lynch a nigger who did or did not rape a certain girl.

I beg the reader's pardon. That was an anachronistic slip. This is a novel about an eighteenth century Negro. Today the black man is everywhere free and equal to the white.[26]

Thus are allusions to the Scottsboro case threaded into the novel. In another contemporary reference, Babouk quiets a spirited discussion about whether white women can produce mulatto children by ruling that only black women can give birth to mulattos. The narrator immediately intercedes:

What nonsense you talk, Babouk!

True enough it is by the millions that the Negro women raped by whites can be counted, but you should know that the terror of terrors is the Negro raper. It is of him that the newspaper headlines scream.

It is in fear of him that the white ladies close their shutters tight and lock their doors; and the white husbands clean their guns.

Beware! The black raper is abroad.

The gorilla!

The monster!

Lynch him! Off with his brown testicles! For everything belongs to the whites. Africa and America! Asia and Europe! And women of all colors and countries.[27]

Speaking sometimes to Babouk, sometimes to the planters, and—most often—to the reader, the narrator rails against the deafness of the master to the complaints of the slave and the inadequacy of official histories that only take into account the experiences of the ruling classes: "for it is not history until the rich man suffers." Behind these complaints, however, looms the gathering storm of black revolution:

> But, Babouk, we have gone beyond your century. Your voice is lost in the past. Your wavering voice is lost in the steaming fold of Saint-Domingue. It is lost in both time and space. And yet it cannot be lost altogether, Babouk. It cannot die in a void. Oh, no. All the wavering voices of the complaining Negro, be they of the dead or of the living, of Africa, or America, yet they will some day be woven into a great net and they will pull that deaf master out of his flowery garden and down into the muddy, stinking field.[28]

Like Langston Hughes in *Scottsboro Limited* and Paul Peters and George Sklar in *Stevedore*, Endore is interested in probing the capacity of blacks to rise up and strike out against centuries of accumulated injustice—even as he recognizes that these struggles will proceed by fits and starts:

> It seemed as if the slaves must be shown a hundred times that nothing but universal revolt could remedy their situation. Degraded by their condition of serfdom, and brutified by their constant heavy labor, they awoke from their slothful apathy only to a brief explosion of rage, which, being unformed and unsupported, was speedily extinguished and, of course, punished in exemplary fashion.[29]

In *Babouk*, the dream of "universal revolt" occurs against the backdrop of the 1789 French Revolution; two years later Babouk inspires the insurrection that will culminate in the Haitian Revolution—even though he will not live to see its success. In the aftermath of a particularly graphic scene in which Babouk mercilessly impales the baby of his former owner's niece and raises its body on his spear as his standard, the narrator offers the following dispassionate commentary:

> Our historians, who always shout reign of terror when a few rich people are being killed and see nothing much worthy of comment when poor are slaughtered by the thousands in the miseries of peace, cry out unanimously: the pen cannot describe the cruelty of these savages!
>
> My pen is not so delicate; it can say, and it will never cease to say: not over a thousand or so of whites were killed in this reign of terror, while the legal and protected slave trade killed over a hundred thousand Negroes a year. Buried its victims in Africa and America and strewed them over the Atlantic Ocean.
>
> One must be exact: the slaves revolted and the reign of terror that had lasted hundreds of years in Saint-Domingue stopped![30]

The final chapter of *Babouk* opens with two juxtaposed epigraphs:

> Divide and rule.
>   —*Policy of the imperialists.*
> Black and white
> Unite and fight!
>   —*Modern rallying cry for world social justice*[31]

In a concluding intervention that would be the cause of great concern among some of Endore's critics on the Left, the narrator steps in to spell out and dramatically underscore the meaning of Babouk's story:

> But beware, whites. Beware!
> Some day the Negro's tomtom will truly be dreadful to hear. It will come out of the forest and beat down the streets of our cities.
> TOM-TOM! TOM-TOM! TOM-TOM!
> And it will hunt us out among the girders of our dying skyscrapers as we once hunted them through their forests.
> TOM-TOM! TOM-TOM! TOM-TOM!
> And white blood will flow, and it will flow for days. It will flow for years. And yet were it to flow for a thousand years, never would it erase the white path of Negro bones that lines the bottom of the ocean from Africa to America.
> But the black man's time must come. Some day it will be his turn to march up and down the world and beat his tomtom and cry out:
> Arise, you black cowards! Arise, you universal boot-blacks! Arise!
> Beat out this new proclamation to the world.
> TOM-TOM! TOM-TOM! TOM-TOM!
> This is the world of men. Black and yellow and red and white. . . .
> This is the world of men and of women and of children.

> This is THE world. The ONLY world. The WORLD of ALL. . . .
>
> Oh, black man, when your turn comes, will you be so generous to us who do not deserve it?[32]

Endore, in short, left no doubt about his belief that blacks would necessarily have to play the central role in achieving their own liberation. The conclusion of *Babouk*, however, seemed ambivalent about how the slogan "Black and White, Unite and Fight!" would be practiced. Langston Hughes's *Scottsboro Limited* made it clear that the will of his characters to struggle for their freedom was inextricably connected to their emerging alliance with the white workers—as did Peters and Sklar's *Stevedore*. John Wexley's *They Shall Not Die* considerably downplayed the issue of the capacity of the black characters for historical agency, placing their fate in the hands of the National Labor Defense lawyers. In sharp contrast, Endore seems to suggest that the relationship of whites to black struggles should be that of sympathetic supporters who embrace the historical and political logic of black nationalist insurgencies.

The handful of critics who reviewed *Babouk* often recoiled sharply before its brazenness, with a *New York Times* reviewer taking sharp exception to Endore's unflattering portrayal of whites in the novel:

> Babouk soon ceases to be clearly individualized, becoming a symbol rather than a person, while the author's delineation of the whites as universally mean, hypocritical, stupid, brutal and generally despicable seems, to put it mildly, just a little bit exaggerated. One feels that even among the white men and women of Saint-Domingue there were probably at least a few ordinarily decent people.[33]

Paul Allen complained that Endore seemed too eager to hammer his readers with his political views:

> Mr. Endore would have done better to let his story speak for itself, leaving the reader to find the horror and shame that pervades it. As it is, the heavy irony and the strident shrieking about the brotherhood of man culminating, on the last two pages of the book, in gibberish and exclamation points, practically ruins the book as either literature or propaganda.[34]

Martha Gruening praised *Babouk's* promise, but chided Endore for failing to achieve it:

> "Babouk" is a horrible and an unforgettable book, but it somehow misses being a great, tragic, or memorable one, despite its important and tragic material and the occasional power and intensity of Mr. Endore's writing. This is all the more disappointing because Mr. Endore's material would lend itself so easily to a magnificent novel of epic proportions—a novel particularly timely in view of the fashionable agrarian nostalgia for the peculiar institution—or, on the other hand, to a soberly and thoroughly

documented history of the Caribbean slave trade. Mr. Endore has written neither.[35]

Virtually none of the reviewers seemed to detect any contemporary overtones in *Babouk*—with the exception of African American critic Eugene Gordon, who offered much praise for the novel in the pages of *New Masses*, but issued some sharp reservations as well, particularly about its concluding polemics:

> Endore confuses his reader toward the end. Whether this weakness in Babouk derives from the author's own confusion I do not know, but I suspect it does. . . . he gives the impression that he is dealing with "race" forces instead of economic forces. He does this by going into a polemic against whites in general by presenting the ghastly picture, and justifying it, of the blacks rising en masse, in their newly organized strength, against the whites in general. Here, he does not touch upon the fact that whites and blacks under capitalism, under imperialism, are divided into classes. . . . Mr. Endore cries to the blacks and whites to unite and fight. Good! Then he suddenly leads them into a situation from which there can come nothing but "race war"; leaves the blacks and the whites glaring murderously at each other across an unsubstantial color line—*all* the blacks glaring murderously at all the whites.[36]

But orthodox Communist opinion probably had little impact on the fate of *Babouk*: it sold four hundred copies in the middle of the Depression, then had its plates destroyed and disappeared from literary history until its reprinting in 1991[37]—before disappearing again.

Like Guy Endore, Arna Bontemps regarded the French and Haitian revolutions as pivotal moments in the history of black revolts in the New World. These interrelated events provide the historical backdrop for his 1936 novel *Black Thunder*—based upon the historic, and aborted, rebellion organized by Gabriel Prosser in Richmond, Virginia in 1800. In a more immediate sense, however, Bontemps's novel was inspired and deeply informed by his anguished and brooding response to the Scottsboro case, which pervaded the daily atmosphere in Huntsville, Alabama, where he lived and worked during the early years of *Black Thunder's* composition.

Born in Alexandria, Louisiana in 1902, Arna Bontemps moved to Los Angeles, California with his family three years later, where they sought to flee the repression of Jim Crow. There he was educated at San Fernando Academy and Pacific Union College—Seventh-Day Adventist schools—graduating in 1923. Like many aspiring writers of his generation, he responded to the siren-call of Harlem, hub of the "New Negro Renaissance," after Jessie Fauset, editor of the NAACP's *Crisis* magazine, accepted one of his poems for publication. Bontemps moved to New York City shortly afterwards, arriving in 1925. He quickly secured a teaching job at a Seventh-Day Adventist school, the Harlem Academy, and in 1926 he married a former student, Alberta Johnson. He also quickly established new friendships with many of the rising young stars of the "New Negro" movement: Countee Cullen, Harold Jackman, and, most notably, Langston Hughes, who would become a lifetime friend and collaborator.

The publication in 1931 of Bontemps's first novel, *God Sends Sunday*, based on the life of a favorite, roguish uncle, coincided with the early years of the Depression and the rather abrupt ending of the "Harlem Renaissance." According to Arnold Rampersad:

> Almost from the start, his literary career and associations placed him in conflict with the authorities at the Harlem Academy, who were wary of the secular and erotic aspects of the Renaissance. The mention of God in the title of his novel was an act of blasphemy, the last straw. In 1931, Bontemps was sent away to Huntsville, Alabama, to teach at another Adventist school, Oakland Junior College.[38]

Bontemps's tenure at Oakwood Junior College, thirty miles from Scottsboro, during the period 1931–34, coincided with the arrests and first trials of the Scottsboro Boys; the 1932 Supreme Court decision, *Powell v. Alabama*, which overturned the original verdict; the second trials in Decatur, Alabama in 1933; and Judge Horton's courageous decision to overturn that verdict. As Bontemps later recalled:

> It was the best of times and the worst of times to run to that state for refuge. Best, because the summer air was so laden with honeysuckle and spiraea, it almost drugged the senses at night. . . . That was also the year of the trials of the nine Scottsboro boys in nearby Decatur. Instead of chasing possums at night and swimming in creeks in the daytime, this bunch of kids without jobs and nothing to do had taken to riding empty box cars. . . . Several times I was tempted to go over and try my luck at getting through the crowds and into the courtroom to hear the trials. My wife did not favor this, and neither did I on second thought. We were not transients, as were most of the spectators, and that made a difference.[39]

After three years of working under the shadow of Scottsboro, however, Bontemps had had enough. His literary habits—particularly his practice of borrowing library books by mail—had made him suspect in the community; and officials at the Seventh-Day Adventist school issued him an ultimatum: "I would have to make a clean break with the unrest in the world as represented by Gandhi's efforts abroad and the Scottsboro protests here at home." He could do so publicly, the school head informed him, "by burning most of the books in my small library, a number of which were trash in his estimation anyway, the rest, race-conscious and provocative. *Harlem Shadows, The Black the Berry, My Bondage and Freedom, Black Majesty, The Souls of Black Folk,* and *The Autobiography of an Ex-Coloured Man* were a few of those indicated."[40] Horrified, his wife pregnant with another child, and facing no prospects for employment, Bontemps and his family abandoned Alabama for his parents' home in California, where he completed *Black Thunder*—and Scottsboro continued to haunt his dreams.

As in Grace Lumpkin's *A Sign for Cain* and Guy Endore's *Babouk,* the parallels between *Black Thunder* and the Scottsboro case are inexact, but, in Bontemps's hands, they are persistent and powerful. First and foremost—like Langston Hughes, Paul Peters and George Sklar, and Guy Endore—Bontemps is concerned with directly addressing the issue of black agency. Gabriel's Rebellion, like the Haitian Revolution, triggered off widespread speculation among whites about the source of the ideas that led blacks to revolt. Behind the assumption that these actions could only have been provoked by [white] "outside agitators" stood, of course, the belief—as widespread in post-Reconstruction popular culture and historiography as it was during the era of the transatlantic slave trade—that Africans and their descendants were unfit to claim their economic, political, or legal freedom—that they were not capable of either imagining their freedom or waging struggle to achieve it. Although no one on either side of the Scottsboro case claimed that the boys had been motivated by the Communists to commit their alleged crimes, the prominent role that Communists played in their defense—in the United States and throughout the world—fueled charges by Alabamians in particular, and many others, including the NAACP, that "outside agitators" were both fomenting racial discontent in the South and undermining serious efforts to secure the freedom of the Scottsboro Boys. Bontemps's sharp awareness of the analogies between these historical moments informed his careful probing of these debates throughout *Black Thunder,* where the issue is succinctly framed in Creuzot's musings:

> Actually, *could* it be? *Could* these tamed things imagine liberty, equality? Of course, he knew about San Domingo, many stories had filtered through, but whether or not the blacks themselves were capable of the divine discontent that turns the mill of destiny was not answered.[41]

The first section of *Black Thunder,* aptly titled "Jacobins," begins with the brutal murder of an old slave, Bundy, at the hands of Gabriel's owner, Thomas Prosser—an act that contributes to the general sullen mood of the slave community. "The blacks are whispering," is the constant refrain of the conversations of the people who gather at the print shop of the Frenchman Creuzot—and a great deal of this section of the novel is concerned with exploring *why* the blacks are whispering. Along the way Bontemps explores the limits of white perceptions about black agency and effectively refutes the widespread myth of black passivity in the face of continual oppression.

In a scene where Gabriel eavesdrops on the conversation between Creuzot and his visitor from Philadelphia, Alexander Biddenhurst, Bontemps makes it plain that the sources of Gabriel's political inspiration rest in the spoken, not the written, word: "But Gabriel . . . was innocent of letters and interested only in the words of the other men."[42]

Creuzot and Biddenhurst invoke the slogan of liberty, equality and fraternity; and discuss the Rights of Man and Thomas Jefferson's *Notes on Virginia*. Their words resonate with Gabriel, but the passage strongly suggests that his views derive from other sources:

> Here were words for things that had been in his mind, things that he didn't know had names. Liberty, equality, frater—it was a strange music, a strange music. And was it true that in another country white men for these things, died for them?[43]

At the same time Bontemps deliberately places anachronistic terms in the mouths of the white radicals—as if to underscore their kinship with Communists. Thus Biddenhurst offers a class analysis of the American South: "Here we have the planter aristocracy. We have the merchants, the poor whites, the free blacks, the slaves—classes, classes, classes. . . . Liberty, equality and fraternity will have to be won for the poor and the weak everywhere if your own revolution is to be permanent. It is for us to awaken the masses."[44] Many 1930s readers undoubtedly would have recognized this statement as a central tenet of the Communist International, particularly during its "Third Period." And they would have certainly understood the contemporary relevance of Creuzot's skepticism about whether "there was hope for the masses in Virginia, that white and black workers, given a torch, could be united in a quest,"[45] or his subsequent references to inciting "the proletariat"—a term that did not come into popular usage until many years after the action of the novel.

Passages such as these effectively emphasize the gulf between the consciousness of Gabriel and his fellow slaves and that of the white radicals in the novel. Gabriel's plot has been shaped by the grim realities of slavery, Bontemps makes clear, and shaped—for good and ill—by oral traditions, the prophetic faith of Afro-Christianity, and the inspirational example of Toussaint L'Ouverture's successful revolution in Haiti; it does not need, he strongly implies, the textual authority provided by European revolutionary documents. The white radicals support black aspirations, to be sure—albeit with varying degrees of enthusiasm—and the fact of the American and French Revolutions clearly provides a significant political and ideological context for the prospects of black rebellion. But the blacks are clearly capable of taking up arms and acting on their own behalf.

Having answered Creuzot's question with a resounding yes, Bontemps casts his eye on the characteristics of the white supporters of black insurrection. His skepticism about white allies—and, therefore, about the cogency of the slogan "Black and White, Unite and Fight!"—is displayed most clearly in his depiction of the white radical Alexander Biddenhurst, an "outside agitator." From his first appearance in *Black Thunder*, Biddenhurst is distinguished by the earnestness of his views and his eloquent mastery of the relevant texts of the world revolutionary movement. His activities, however, seem to revolve around conversations in the comfort of Creuzot's house and shop about the prospects of a black uprising, and alcohol-induced exchanges with the local citizenry about equality. Warned by his sometimes consort Melissa, a free "mulattress," Biddenhurst—in the name of prudence—flees to Philadelphia on the eve of Gabriel's Rebellion, mouthing as he leaves the platitudes of his revolutionary faith:

> The revolution of the American proletariat would soon be something more than an idle dream. Soon the poor, the despised of the earth, would join hands around the globe; there would be no more serfs, no more planters, no more classes, no more slaves, only men.
> The stagecoach lurched; dust poured in the open windows.[46]

There is the faintest hint that Biddenhurst, who shortly will be followed to Philadelphia by Creuzot and his family—like the contemporary white supporters of the Scottsboro Boys—enjoys the freedom to travel to the North, where he can have the vicarious experience of black revolt through newspaper reports and other secondhand accounts, and suffer none of the grim consequences. Nevertheless, Bontemps is not completely unsympathetic. He recognizes, for example, the feelings of helplessness and futility that someone like Biddenhurst experiences, and he acknowledges his important role as a *supporter* of black liberation:

> The only thing left for him was to continue his same endeavors. There were many tender old people (with time and money to burn) who longed to see justice triumph, even if it would reduce their own fortunes. More and more these would be willing to support the small, semi-secret groups working for the deliverance of bondsmen here and there, singly, as the occasion arose. . . .
> And young men like Alexander Biddenhurst, lawyers, scholars, poets, would receive their support in the discharge of this work while neglecting the saner courses of business. They would keep the spark alive by agitating, agitating, agitating. They would work in the schools, winning the youngsters and the teachers; they would go among the blacks, flaunt the old taboos, slap the hands of the wretches, tell them there was deliverance ahead and to be ready for the revolution at any hour. Comforted by the philosophers and writers, they would carry on their near-hopeless mission; they would be the drip of spring on the determined rock.[47]

Bontemps's ambivalent portrayal of Biddenhurst and other white radicals in *Black Thunder* not only captured the tensions of his ongoing struggle with his Seventh-Day Adventist background, it probably reflected the perspective of many blacks—artists, intellectuals, and workers—who were drawn by the rhetorical appeal of the Communist Party at the same time that they expressed skepticism about its utopian hopes.

Like other authors of the dominant Scottsboro Narrative of the 1930s, Bontemps was prepared to tackle the nexus of race and sex that was so deeply embedded in the Scottsboro case, even if he did so obliquely. As the vengeful slave Criddle lies in wait outside of his assigned post—the home of a poor white—on the evening of the planned rebellion, Bontemps suddenly introduces the character of Grisselda, the farmer's daughter, who is described in sexually suggestive terms:

> Grisselda remembered that she was sixteen now; she knew why a young farmer had noticed the color of her hair and why the middle-aged store-keeper had recently pinched her cheek. A lewd thought dropped into her thought like a pebble falling into a well.[48]

Meanwhile, Criddle lurks outside in the dark and the rain:

> He passed a thumb meditatively down the edge of his scythe-sword and derived an unaccountable pleasure from the thought of thrusting it through the pale young female that stood looking in the darkness with such a disturbed face. . . . He held his sword arm tense; the scythe blade rose, stiffened and remained erect.[49]

Bontemps seems to momentarily flirt with the idea of sexual violence as a legitimate form of racial revenge, before he retracts the threat—even as the language of the following passage retains the power of sexual suggestiveness:

> It was like hog-killing day to Criddle. He knew the feel of warm blood, and he knew his own mind. He knew, as well, that his scythe-sword was ready to drink. He could feel the thing getting stiffer and stiffer in his hand. . . .
>
> And that there gal in the long night-gown and the lamp in her hand. Humph! She don't know nothing. Squeeling and a-hollering around here like something another on fire. She need a big buck nigger to—no, not that. Gabriel done say too many times don't touch no womens. This here is all business this night. What that they calls it? Freedom? Yes, that's the ticket, and I reckon it feel mighty good, too.
>
> Where the nation that gal go to? . . . Where she go?[50]

As a result of her close call with Criddle—with all of its overtones of the threat of sexual violence—Grisselda, rather like the accusers of the Scottsboro

Boys, Ruby Bates and Victoria Price, is somewhat elevated in her social status among the white townspeople. She reappears—"the dazed, vacuous girl whose parent got himself impaled on Criddle's scythe-sword"—during the trials and summary executions of the "anonymous Negroes" held responsible for the failed rebellion: "the girl, waiting hungrily for each new kill, neither blinked nor recovered. She stared, leaning a little forward, tearing her homespuns absently. There seemed to be an impression among those who had ceased to regard her presence that she was entitled to whatever compensation she could wring from the sights."[51]

Bontemps thus deflects the idea that Gabriel's rebellion carries with it any hint of sexual transgression at the same time that he makes plain (as Endore does in *Babouk*) that the real sexual crimes, in the context of slavery and in the contemporary Jim Crow South, have been committed by white men against black women—as when he dramatizes the sexually charged whipping that Thomas Prosser administers to Juba, Gabriel's sweetheart, towards the end of the novel:

"H'ist them clothes."

She understood and obeyed, snatching the old tattered skirt up over her naked buttocks. . . .

Here's something else to toss up your petticoats for, ma'm. Here's something worth fluttering you hips *about*. Understand?

She didn't speak, didn't even flinch. Presently her thighs were raw like cut beef and bloody. Once or twice she turned her head and threw a swift, hateful glance at the powerful man pouring the hot melted lead on her flesh, but she didn't cry out or shrink away. The end of the lash became wet and began making words like *sa-lack, sa-lack, sa-lack* as it twined around her thin hips. *Sa-lack, sa-lack, sa-lack*.[52]

In *Black Thunder* Bontemps also directly confronts the unasked question that often hovered over the historiography of slavery, particularly in the United States: how to account for the apparent acquiescence of the black community in its own degradation; or, to put it more baldly, how to account for its active complicity in its own oppression? In contemporary terms, the analogous question—and the one that Bontemps's friend Langston Hughes had raised so loudly and vehemently during his southern reading tour—was, how to account for the apparent silence of Alabama blacks in particular, and southern blacks in general, about the plight of the Scottsboro Boys? Bontemps grappled with these questions through his meticulous depiction of the intricate social structure and psychology of the slave community, particularly the characters who betray Gabriel's plans: Ben and Pharaoh.

The crucible of slavery in the United States gave birth to a wide range of distinct personalities, representative in a broad sense of strikingly different modes of coping with racial oppression, and Bontemps was certainly not alone in drawing upon this repertory of historical types—and stock literary figures—in his fiction. In his portrayal of Ben and Pharaoh, however, Bontemps seems particularly bent on revealing the factors that led these characters to pursue the paths they chose—and on bringing their actions to judgment.

Based on the historical figure of Ben Woolfolk, an active participant and recruiter in Gabriel's plot, Ben in *Black Thunder* is a comfortable, self-satisfied house servant who lives for his obsequious devotion to his master Mosely Sheppard. He reluctantly takes the vow of silence about Gabriel's conspiracy out of a sense of loyalty to Bundy, who had pleaded with him to listen to Gabriel right before he was killed. Once he learns the details of the plot, however, he breaks down in terror:

> "Oh, Lord Jesus," he said, crying.
>
> A powerful elbow punched his ribs, and Ben raised his head without opening his eyes.
>
> "What's the matter, nigger, don't you want to be free?"
>
> Ben stopped sobbing, thought for a long moment.
>
> "I don't know," he said.[53]

Ben does *not* want to be free; he is haunted by the threat of the destruction of the world he knows, and values:

> Ben was tortured with the vision of filthy black slaves coming suddenly through those windows, pikes and cutlasses in their hands, their eyes burning with murderous passion and their feet dripping mud from the swamp. He saw the lovely hangings crash, the furniture reel and topple, and he saw the increasing black host storm the stairway. In another moment there were quick, choked cries of the dying, followed by wild, jungle laughter. Then it occurred to Ben which side he was on.[54]

Ben's complete subservience to his master's interests is paralleled by Pharaoh's utter pettiness. Pharaoh's main complaint about Gabriel is that he has gotten "biggity," but his character is that of a chronic malcontent who has no qualms about elevating his personal complaints above the needs of the black community. Needless to say, Ben and Pharaoh's betrayal of Gabriel's plans are the pivot of dramatic action in *Black Thunder*. In his extended meditation on the destructive role that black traitors have played during the long, fitful march towards African American freedom, Bontemps could well have been mindful of ominous parallels with the Scottsboro case. He would have certainly been aware that one factor—albeit a relatively minor one in the overall scheme of things—was the reprehensible willingness of two of the Scottsboro Boys—Clarence Norris and Roy Wright—to offer false testimony to the authorities in a vain attempt to secure their freedom. Such behavior, Bontemps warned in *Black Thunder*, had dire consequences, and he clearly took devilish glee in meting out the appropriate punishment to those blacks who betrayed the cause.

Finally, in the last section of *Black Thunder* Bontemps graphically illustrates how the "state apparatus" springs into action in the aftermath of the threat of black rebellion, as government officials, the press, and the courts seek to restore public confidence in the stability of the state. A series of sharply defined episodes dramatize the paranoia that occurs when blacks step outside of their socially and historically circumscribed roles—and the reign of terror that is unleashed against the black community as a result. Scene after scene registers outbreaks of hysteria among the white populace, followed by vengeful acts of white retribution: "Half a dozen obscure blacks had just been hanged following a summary hearing before the wigged justices who composed the *Oyer and Terminer* tribunal for the emergency. The stern show of retribution, as had been expected, was proving itself a useful sedative, but the city was still mad, still frothing at the mouth."[55] In such a climate, no distinction is made between the guilty and the innocent, and the mob rules: "Militiamen still milled the streets . . . periodically dragging anonymous Negroes before the justices,"[56] many of them "inconspicuous nobodies"—like, needless to say, the Scottsboro Boys.

Given the inflamed public passions aroused by Gabriel's thwarted rebellion, there is simply no possibility of justice for him and his allies—even though the novel confers upon them great poise and dignity as they face their inevitable deaths. The actual circumstances of the historical episode Bontemps had chosen as the basis of *Black Thunder* precluded any other conclusion. Nor was Bontemps sanguine about the prospects of the Scottsboro Boys escaping the death penalty—the outcome of their case was far from clear in the mid-1930s. This may account for the grey and melancholy mood that pervades the end of *Black Thunder*; in the final analysis, Bontemps's imagination could go no further than historical facts, and his contemporary circumstances, would permit.

As in their responses to Guy Endore's *Babouk*, most critics read *Black Thunder* as historical fiction, devoid of any contemporary implications, and generally praised Bontemps for his fidelity to the historical record—as Martha Gruening did in the *New Republic*:

> Mr. Bontemps' recreation of this historical incident is at once faithful and imaginative. . . . Writing with great restraint he conveys in a deeply moving manner the tragic dignity of his subject.[57]

Similarly, the *Saturday Review of Literature* noted that Bontemps

> has brought alive the Richmond of that month of terror, brought it alive to move convincingly and swiftly in the pages of a fiction which is a good deal truer than most history as historians used to write it before fictionists began to write it for them.[58]

And the Lucy Tompkins, writing in the *New York Times*, praised *Black Thunder* for the beauty of its prose, rising to flights of oratory of her own:

> If one were looking for a sort of prose spiritual on the Negroes themselves, quite aside from the universal dream that they bear in this story, one could not find it more movingly sung. And the whole flows in rhythms coincident with the emotional patterns. For readers who will read far enough beyond page one to get the feel of the thing, the simplicity, precision and elasticity of the prose alone should be a happy reward.[59]

In the final analysis, however, none of these three novels—not *A Sign for Cain*, not *Babouk*, not *Black Thunder*—had the kind of immediate impact upon contemporary audiences that their creators hoped for or their critics feared. Still, they exist as an important "archive" of the connections between the version of the Scottsboro case promoted by the Communist Party and the International Labor Defense and novelists whose imaginations were deeply influenced by their relationships to the literary Left.

CHAPTER FIVE

RICHARD WRIGHT'S SCOTTSBORO
OF THE IMAGINATION

> "The Reds sure scared them white folks down South when they put up
> that fight for the Scottsboro boys."
>
> *— Lawd Today!*[1]

> "Oh, Bigger," said Mr. Dalton.
> "Yessuh."
> "I want you to know why I'm hiring you."
> "Yessuh."
> "You see, Bigger, I'm a supporter of the National Association for
> the Advancement of Colored People. Did you ever hear of that
> organization?"
> "Nawsuh."
> "Well, it doesn't matter," said Mr. Dalton. . . .
>
> *— Native Son*[2]

> "Where's your father?"
> "Dead."
> "How long ago was that?"
> "He got killed in a riot when I was a kid—in the South."
> There was silence. The rum was helping Bigger.
> "And what was done about it?" Jan asked.
> "Nothing, far as I know."
> "How do you feel about it?"
> "I don't know."
> "Listen, Bigger, that's what we want to *stop*. That's what we Communists
> are fighting. We want to stop people from treating others that way. I'm
> a member of the Party. Mary sympathizes. Don't you think if we got
> together we could stop things like that?"
> "I don't know," Bigger said; he was feeling the rum rising to his head.
> "There's a lot of white people in the world."
> "You've read about the Scottsboro boys?"
> "I heard about 'em."
> "Don't you think we did a good job in helping to keep 'em from killing
> those boys?"
> "It was all right."
>
> *— Native Son*[3]

Any Negro who has lived in the North or the South knows that times without number he has heard of some Negro boy being picked up and carted off to jail and charged with "rape." This thing happens so often that to my mind it had become a representative symbol of the Negro's uncertain position in America. Never for a second was I in doubt about what kind of social reality or dramatic situation I'd put Bigger in, what kind of test-tube life I'd set up to evoke his deepest reactions. Life had made the plot over and over again, to the extent that I knew it by heart.

— "How Bigger Was Born"[4]

EVEN AS THE SCOTTSBORO CASE BEGAN to drift in and out of public consciousness after 1937, when the state of Alabama decided to drop the rape charges against four of the defendants—Willie Roberson, Olen Montgomery, Eugene Williams, and Roy Wright—it was becoming reconstituted in the imaginative landscape of a new and powerful voice in African American writing represented by Richard Wright. It is no accident that the emergence of Richard Wright coincided almost exactly with the period of heightened public concern about the Scottsboro case, for no African American writer before him had captured more effectively the devastating social and psychological terror unleashed on black men—in part by the reckless or calculated accusations of rape. For Wright, the Scottsboro case was epiphenomenal, simply the most recent—and dramatic—manifestation of the injustice routinely inflicted upon blacks in the Jim Crow South. Unlike the steady procession of northern-based writers, intellectuals, and political activists who made the trek to Alabama to bear witness to the ordeal of the Scottsboro Boys during the 1930s—and who carried their skirmishes and confrontation with southern authorities and racists as badges of honor—Richard Wright encountered Scottsboro (the case broke during his fourth year in Chicago, about three years before he joined the Communist Party) through the deeply embedded structures of thought, feeling, and memory that had been shaped by his experiences as a black man who had come of age in the Jim Crow South during the 1920s.

A refugee from the economic and spiritual poverty and the crippling social and political restrictions of the Jim Crow Mississippi where he had been born to an illiterate sharecropper and a schoolteacher in 1908, Wright had lived a peripatetic existence with his mother and younger brother, traversing Natchez, Mississippi; Memphis, Tennessee; Jackson, Mississippi; Elaine, Arkansas; Greenwood, Mississippi; and Jackson before fleeing the South for the promise of a better life in Chicago in 1927.

Armed with a ninth-grade Mississippi education, Wright worked at various jobs during his early years in Chicago before securing a position at the post office. Throughout these years Wright pursued reading with the voracious appetite for reading that had consumed him since his graduation from high school; at the same time he doggedly pursued his own fugitive literary efforts in his spare time—a continual source of tension between him and his family. He also began associating with a small group of white friends from the post office, including Abraham Aaron, a Communist who shared Wright's literary ambitions. In the late fall of 1933 Aaron urged Wright to come to a meeting of the Chicago John Reed Club, one of proletarian writers and artists organized by the Communist Party at a national conference in Chicago in May 1932—and one explicitly committed to political agitation and propaganda, including

> holding public lectures and debates, establishing JRC art schools and schools for worker-correspondents, active participation in strikes and demonstrations, making of posters and contributing of literature to working class organizations, giving of chalk talks, dramatic skits and other entertainment at workers' meetings, and active assistance to campaigns (Scottsboro, Mooney, Berkman) involving special issues.[5]

Wright was enthusiastically welcomed by the white membership of the club and he quickly reciprocated. Soon he was completely immersed in its cultural and political milieu—the world of proletarian internationalism where the Scottsboro case figured prominently. As he later recalled:

> It was not the economics of Communism, nor the great power of trade unions, nor the excitement of underground politics that claimed me; my attention was caught by the singularity of the experience of workers in other lands, by the possibility of uniting scattered but kindred peoples into a whole. It seemed to me that here at last, in the realm of revolutionary expression, Negro experience could find a home, a functioning value and role.[6]

The Chicago John Reed Club was Richard Wright's university, as his most recent biographer Hazel Rowley has aptly pointed out;[7] it was also his introduction to the world of internecine Party politics. Elected as Executive Secretary of the Club as a compromise between contending factions in the membership a mere several months after joining, Wright subsequently joined the Communist Party in early 1934 to affirm his relationship to a group he now regarded as vital to his existence. From that time until his official rupture with the Communist Party, signaled by the publication of his essay "I Tried to be a Communist" in the July and August 1944 issues of *Atlantic Monthly*, Wright moved and worked within the orbit of the Party.

Through Party circles Wright published his fledgling literary efforts in such journals as *Left Front, Anvil, New Masses, Midland Left, International*

Literature, and *Partisan Review;* and "Between the World and Me" was re-printed in the classic anthology *Proletarian Literature in the United States.* He leaped into an active role in the national congress of John Reed Clubs in 1934 and took the floor to argue strenuously against the decision by the Party to dissolve the clubs. Wright hitchhiked from Chicago for his first trip to New York to attend the first American Writers' Congress. Back in Chicago he met Langston Hughes—an admirer of Wright's ever since he had read his poem "I Have Seen Black Hands" in the *New Masses* in 1934—and Hughes's close friend and collaborator Arna Bontemps, who had recently moved with his family to Chicago and whose novel *Black Thunder* Wright would review in April 1936. He was hired by the Illinois unit of the Federal Writers' Project to research and write the history of ethnic and working class communities in Chicago and Illinois. As supervisor of the project—the first African American in the region to be appointed to such a position—Wright took particular care to ensure that African American life and history was taken seriously and inte-grated into the history. He also became a central figure in the South Side Writers' Group, a group of young black Chicago writers that included Arna Bontemps, Frank Marshall Davis, Theodore Ward, Margaret Walker, Marion Minus, Fenton Johnson, Ed Bland, and others—a gathering that provided a sounding board for his provocative and influential 1937 essay "Blueprint for Negro Writing," extolling the liberatory potential of a Marxist perspective for new directions in African American literature.[8]

In 1937, Wright took a decisive and fateful step by moving to New York City to pursue his career as a writer. Although he had drifted from the Party in Chicago—a rift that he dramatizes in his account of his physical expulsion from the Chicago's 1937 May Day parade in the final passages of the second part of his autobiography *American Hunger*—Wright immersed himself in Party circles in New York City, where he became Harlem editor for the *Daily Worker*—churning out forty signed articles and numerous briefer, unsigned dispatches over the next year, including coverage of the continuing saga of the Scottsboro Boys. As Wright biographer Michel Fabre notes:

> he was forced to familiarize himself with the major problems of the day, and in particular with the intricacies of the famous Scottsboro case. . . . The Harlem bureau mentioned it at least once a week, either to announce a judicial victory, or to invite its readers to attend a demonstration or con-tribute funds. If Wright later alludes to the "Scottsboro Boys" in his books, it is because he became so involved with the case at this time.[9]

While Wright's social perspective was shaped by his sustained engagement with a political culture defined by the Communist Party in particular and the Left in general, he brought to this arena a sensibility already permanently formed—some would say "scarred"—by his upbringing in the Jim Crow South. And at the center of this sensibility stood the terror provoked by the accusation of rape—an issue to which Wright returned over and over again over the course of his long career, from his earliest published stories in *Uncle Tom's Children* through his blockbuster novel *Native Son* to the last novel published in his lifetime, *The Long Dream,* and his posthumously published fiction. For Wright, the Scottsboro case was grafted upon the deep structure of personal and cultural memory, providing a point of departure for a probing and sustained critique of Jim Crow terror and racial injustice and subordination in the United States.

From the very beginning of his literary career, Wright's work was distinguished by its preoccupation with the horror of lynching. In one of the early poems that brought him to the attention of readers on the Left, published in the July–August 1935 issue of *Partisan Review,* "Between the World and Me," the narrator stumbles upon the aftermath of a recent lynching in a grassy clearing and immediately begins to imaginatively recreate it:

> There was a design of white bones slumbering forgottenly upon a cushion of ashes.
> There was a charred stump of a sapling pointing a blunt finger accusingly at the sky.
> There were torn tree limbs, tiny veins of burnt leaves, and a scorched coil of greasy hemp;
> A vacant shoe, an empty tie, a ripped shirt, a lonely hat, and a pair of trousers stiff with black blood.
> And upon the trampled grass were buttons, dead matches, butt-ends of cigars and cigarettes, peanut shells, a drained gin-flask, and a whore's lipstick. . . .[10]

The poem achieves its effect through the relentless accumulation of concrete details, and the reference to "a whore's lipstick" is particularly striking, suggesting as it does not only the presence of women at the scene, but disreputable ones at that; it would not require much of a stretch of the imagination to conceive of the presence of a Victoria Price or—before her political rehabilitation—Ruby Bates at this spectacle.

The narrator assumes the position of both spectator and victim, staring at the way "the sun poured yellow surprise into the / eye sockets of a stony skull," while his mind is "frozen with cold pity for the life / that was gone" and his heart is "circled by icy walls / of fear." He is then propelled into the experience of the lynch victim, graphically reliving the excruciating details of his death:

> And then they had me, stripped me, battering my teeth into my throat till I swallowed my own blood.
> My voice was drowned in the roar of their voices, and my black wet body slipped and rolled in their hands as they bound me to the sapling.

And my skin clung to the bubbling hot tar, falling from me in limp
patches.
And the down and quills of the white feathers sank into my raw flesh,
and I moaned in my agony.
Then my blood was cooled mercifully, cooled by a baptism of
gasoline.
And in a blaze of red I leaped to the sky as pain rose like water, boil-
ing my limbs. . . .

In his complete identification of the narrator and victim in "Between the
World and Me," Richard Wright had discovered his subject. The experience
of lynching isolated him from the (white) world at the same time that it conferred upon him the insight and cultural authority to interpret it to that very
same world.

At the outset of his literary career, Wright saw himself as a bridge between
an African American world whose terms of existence he intimately understood and the universe of his white readers on the Left who—however well
meaning—often failed to understand black life or who were sometimes
blinded to its realities by ideological formulations that failed to grasp its complexity. "The Communists," he wrote, ". . . had oversimplified the experience
of those whom they sought to lead":

In their efforts to recruit masses, they had missed the meaning of the lives
of the masses, had conceived of people in too abstract a manner. I would
try to put some of that meaning back. I would tell Communists how common people felt, and I would tell common people of the self-sacrifice of
Communists who strove for unity among them.[11]

He initially sought to fulfill this ambition through a series of biographical
sketches of some of the black Communists he had come to know in Chicago,
like David Poindexter, a firebrand leader and speaker, whose freewheeling
style would lead to his trial as "an enemy of the working class," and whose
eyewitness account of the actions of a lynch mob in Nashville, Tennessee
provided the basis of Wright's short story "Big Boy Leaves Home." Wright
tentatively titled this project *Heroes, Red and Black*, but later abandoned it as
the stories that would become collected in *Uncle Tom's Children* increasingly
claimed his attention.

Of the stories that appeared in this groundbreaking work of thematically
related stories—arguably Wright's best—two in particular reflect Wright's attempt to bridge the worlds of southern African American life and Communism: "Big Boy Leaves Home" and "Bright and Morning Star."

In "Big Boy Leaves Home," a southern idyll abruptly turns into tragedy.
Four black young men play hooky from school, take to the woods, sing, chant
rhymes, play the dozens, wrestle with each other, and cross the line to take a
dip in old man Harvey's swimming hole—territory that has been explicitly

banned to black people. They frolic nakedly—and innocently—in a scene replete with edenic imagery. A white woman suddenly intrudes upon the scene and, startled, backs away from the naked boys in a way that prevents them from retrieving their clothes. Big Boy dares to climb over the embankment in her direction. The woman screams. A white man, armed with a rifle and wearing an army uniform, suddenly appears, shooting and killing two of Big Boy's friends, Lester and Buck. He then takes aim at Big Boy's remaining companion, Bobo, and Big Boy wrestles with him for control of the rifle. Big Boy kills the man, and the two boys flee.

In two sharply etched episodes, Wright propels his readers from a classical depiction of American male innocence into a world of gothic, southern terror that pits the innocence of the four black young men against white southern malice, buttressed by the insidious sexual mythology of the black rapist. In this universe, the accusation of rape equals death to the accused, as Big Boy, Bobo, their families, and the entire black community know only too well; and the rest of "Big Boy Leaves Home" is devoted to working out the inevitable consequences of Big Boy's actions—which have triggered off a course of events as implacable as the action of a Greek tragedy. When Big Boy stumbles home to report what has happened, everyone in his family immediately understands what has happened, what will happen, and what must be done:

> His father came into the doorway. He stared at Big Boy, then at his wife.
> "Whut's Big Boy inter now?" he asked sternly.
> "Saul, Big Boy's done gone n got inter trouble wid the white folks."
> The old man's mouth dropped, and he looked from one to the other.
> "Saul, we gotta git im erway from here."
> "Open yo mouth n talk! Whut yuh been doin?" The old man gripped Big Boy's shoulders and peered at the scratches on his face.
> Me n Lester n Buck n Bobo wuz out on ol man Harveys place swimming. . . ."
> "Saul, it's a *white* woman!"
> Big Boy winced. The old man compressed his lips and stared at his wife. Lucy gaped at her brother as though she had never seen him before.[12]

The hastily convened and panicked discussion that follows between Big Boy's family and the church leaders makes it plain that the black community is utterly helpless in the face of inevitable white violence. The only recourse is for Big Boy to flee—and to hope for the best. Everyone tacitly understands, however, that the blood lust of the white community must be satiated.

Big Boy spends the remainder of the story hidden in an abandoned kiln—from which he witnesses from far the final action of the story: the ritual lynching of Bobo, a scene Wright describes in graphic, sexually charged terms:

> They had started the song again:
> *"We'll hang ever nigger t a sour apple tree. . ."*

There were women singing now. Their voices made the song round and full. Song waves rolled over the top of pine trees. The sky sagged low, heavy with clouds. Wind was rising. . . .

Big Boy could see the barrel surrounded by flames. . . .

He smelt the scent of tar, faint at first, then stronger. The wind brought it full into his face, then blew it away. His eyes burned and he rubbed them with his knuckles. He sneezed.

"LES GIT SOURVINEERS!. . . ."

"HURRY UP N BURN THE NIGGER FO IT RAINS!. . . ."

Big Boy saw the mob fall back, leaving a small knot of men about the fire. Then, for the first time, he had a full glimpse of Bobo. A black body flashed in the light. Bobo was struggling, twisting; they were binding his arms and legs.

When he saw them tilt the barrel he stiffened. A scream quivered. He knew the tar was on Bobo. The mob fell back. He saw a tar-drenched body glistening and turning. . . .

There was a sudden quiet. Then he shrank violently as the wind carried, like a flurry of snow, a widening spiral of white feathers into the night. The flame leaped tall as the trees. The scream came again. . . . Then he saw a writhing white mass cradled in yellow flame, and heard screams, one on top of another, each shriller and shorter than the last. The mob was quiet now, standing still, looking up the slopes at the writhing white mass gradually growing black, growing black in a cradle of yellow flame. . . .[13]

Exhausted by the ordeal he has witnessed and completely traumatized, Big Boy flees north with the assistance of a truck driver who has been enlisted to help him escape.

In *Uncle Tom's Children*, stories like "Big Boy Leaves Home," "Down by the Riverside," and "Long Black Song" graphically portray the grim realities of life in the Jim Crow South, probing the question, in Wright's terms: "what quality of will must a Negro possess to live and die in a country that denied his humanity?"[14] These stories pit the individualism of Wright's main characters against escalating patterns of violence designed to break their will and cripple their spirit, dramatizing along the way the futility of individual acts of rebellion and resistance against the racial status quo. If the first three stories in the collection portray the "problem," the last story in the original collection, "Fire and Cloud," propose a "solution."

Unlike the other stories in *Uncle Tom's Children*, "Fire and Cloud" has a topical, Depression-era setting. For the first time Communists appear in the beleaguered southern black community, and the story explicitly grapples with issues of political ideology, strategy, and tactics.

Reverend Dan Taylor faces the greatest crisis of his life. Widely regarded as the leader of the black community, Taylor has been unable to persuade the white authorities to provide any relief for his hungry people—yet they fully expect him to contain the frustrations of the community as the price of his leadership. The religious faith that has sustained him up until this point is strained almost to the breaking point, and his sense of himself as a God-inspired leader is shaken:

> God had called him to preach His word, to spread it to the four corners of the earth, to save His black people. And he had obeyed God and had built a church on a rock which the very gates of Hell could not prevail against. Yes, he had been like Moses, leading his people out of the wilderness into the Promised Land. . . . He had called him to preach his Word, to save His black people, to lead them, to guide them, to be a shepherd to His flock. But now the whole thing was giving way, crumbling in his hands, right before his eyes.[15]

Hadley and Green, two Communist organizers—one white, one black—have been pressuring Taylor to endorse a mass demonstration of blacks and whites to demand relief, but Taylor has resisted them. In the meantime, he faces threats to his leadership of his church from a scheming and conniving Deacon Smith; and faces the challenge of corralling the passions of his fire-brand son, Jimmy, who is prepared to organize his friends for violent resistance against white oppression. All of these contending forces converge at the same time, with the Mayor, the Chief of Police, and the head of the Industrial Squad in the parlor of his house; Deacon Smith and the other deacons in the basement; and the Communist organizers in the Bible room.

Taylor's exchanges with each of these groups offer a clinical view of the alternatives available to the black community under these conditions. His exchange with Hadley and Green is particularly telling because it pits the inadequacy of traditional Afro-Christian responses against the potential effectiveness of the techniques of mass mobilization favored by the Communists:

> "Ahm doin mah duty as Gawd lets me see it," said Taylor.
> "All right, Reverend," said Hadley. "Here's what happened: Weve covered the city with fifteen thousand leaflets. Weve contacted every organization we could think of, black and white. In other words, weve done all *we* could. The rest depends on the leaders of each group. If we had their active endorsement, none of us would have to worry about a crowd tomorrow. And if we had a crowd we would not have to worry about the police. If they see the whole crowd turning out, they'll not start any trouble. Now, youre known. White and black in this town respect you. If you let us send out another leaflet with your name on it calling for. . . ."[16]

At this point in the story Taylor teeters between the obsequious position often conferred upon the stock figure of the traditional black preacher that often appears in proletarian literature and a more dynamic leadership role—but he is not yet prepared to make the leap. His encounter with the Mayor and the other white authorities unfolds along much more predictable lines. Through a combination of ingratiating personal appeals and naked threats, they make their expectations crystal clear:

> "Dan, here we all are, living in good old Dixie. There are twenty-five thousand people in this town. Ten thousand of those people are black, Dan. They're your people. Now, its our job to keep order among the whites, and we would like to think of you as being a responsible man among the blacks. Lets get together, Dan. You know these black people better than we do. We want to feel we can depend on you. Why don't you look at this thing the right way? You know Ill never turn you down if you do the right thing. . . ."[17]

During his meeting with the deacons, Taylor has a heated exchange with Deacon Smith, who, malicious though he may be, accurately describes the truth of his situation:

> "Yeah yuhll march! But yuhs scared t let em use yo name! What kinda leader is yuh? Ef yuhs gonna ack a fool n be a Red, then how come yuh wont come on out n say so sos we kin all hear it? Naw, yuh ain man ernuff t say whut yuh is! Yuh wanna stan in wid the white folks! Yuh wanna stan in wid the Reds! Yuh wanna stan in wid the congregation! Yuh wanna stan in wid the Deacon Board! Yuh wanna stan in wid *every*body n yuh stan in wid *no*body!"[18]

Immediately afterwards, the white authorities carry out their threats by arranging for Taylor's abduction and brutal whipping. As a result Taylor undergoes a dramatic conversion that places him decisively on the side of the people.

Taylor's conversion, Wright takes pains to dramatize, occurs within the terms of his life as he understands and experiences them. In the following conversation with his son Jimmy, Taylor continues to hold his distance from the Communists—who have largely disappeared from the action of the story anyway:

> "We gotta git wid the *people*, son. Too long we done tried t do this thing our way n when we failed we wanted t run out n pay-ff the white folks. Then they kill us up like flies. It's the *people*, son! Wes too much erlone this way! Wes los when wes erlone! Wes gotta be wid our folks. . . ."
>
> "But theys killin us!"
>
> "N they'll keep on killin us less we learn how t fight! Son, it's the people we mus gid wid us! Wes empty n weak this way! The reason we cant do nothing is cause wes so much erlone. . . ."
>
> "Them Reds wuz right," said Jimmy.
>
> "Ah dunno," said Taylor. "But let nothing come between yuh n yo people. Even the Reds cant do nothin ef yuh lose yo people. . . ."[19]

Although the triumphant mass demonstration of blacks and whites that concludes "Fire and Cloud" seems to echo the familiar conclusion of many such proletarian fictions, Wright, in fact, treats the encounter between blacks and Communists with considerable subtlety and complexity. In his choice of a black minister—a figure often routinely treated in a dismissive and contemptuous fashion in the works of his proletarian contemporaries—as the center of consciousness for this story, Wright affirmed an argument that he had advanced in "Blueprint for Negro Writing"—that the politically engaged black writer must, in effect, mine the implications of African American folk life—one important strand of which is deeply embedded in the black church—for its revolutionary potential.[20] Moreover, in the one sustained conversation that occurs between Hadley and Green, Wright carefully distinguishes between the attitudes of the white Communist, Hadley, and the black one, Green, towards Taylor. Hadley is brusque and tends to bully Taylor in a manner not unlike that of the Mayor and his associates; Green seems to be more sympathetic to the nuances of Taylor's situation, more conciliatory in his appeals. Wright seems to suggest here that the Communists have more to learn about the values and beliefs of the people they seek to organize before they can become more effective.

Wright makes it clear, too, that Reverend Taylor's conversion is going to depend not upon his willingness to completely relinquish his deep-seated religious beliefs, but upon his ability to make meaningful connections between his faith and the political exigencies he faces. In a manner that dramatically anticipates the ways in which the modern Civil Rights Movement transformed the themes of traditional African American religious beliefs and songs into effective weapons in the struggle against racial injustice, Taylor discovers the revolutionary potential of Afro-Christian belief:

A baptism of clean joy swept over Taylor. He kept his eyes on the sea of black and white faces. The song swelled louder and vibrated through him. This is the way! He thought. Gawd ain no lie! He ain no lie! His eyes grew wet with tears, blurring his vision: the sky trembled; the buildings wavered as if about to topple; and the earth shook. . . . He mumbled out loud, exultingly: *"Freedom belongs t the strong!"*[21]

In his exploration of the transformation of Reverend Taylor's consciousness in "Fire and Cloud," Wright announced his entry into the ongoing discussion about Communist strategy and tactics during the Scottsboro era. Like Arna Bontemps's *Black Thunder*—only much more directly—Wright's "Fire and Cloud" confronts the issue of African American agency—with or without white allies—in the face of racial oppression.

Wright's political designs become even clearer in the story that emerged, in effect, as the alternative ending to *Uncle Tom's Children*: "Bright and Morning Star." Wright had completed the story in time for it to be included with the triumphant publication of *Uncle Tom's Children* in March 1938, but reconsidered. According to Michel Fabre, Wright was still mulling over its final shape when he received word that he had received the first prize of $500 for *Uncle Tom's Children* in a contest sponsored by *Story* magazine for writers in the Federal Writers' Project—and assurances of a publishing contract. Whit Burnett, one of the editors of *Story*, advised Wright to cut all references in the story to the Scottsboro trial and the International Labor Defense, among other changes, and Wright willingly complied.[22] *Uncle Tom's Children* was greeted with critical accolades; *New Masses* published "Bright and Morning Star" as a special literary supplement in early May, 1938, and it appeared as the concluding story of the collection in subsequent editions.

Like "Fire and Cloud," "Bright and Morning Star" is set in the rural South during the Depression. The consciousness of Aunt Sue, the central character in "Bright and Morning Star," has been molded by religious beliefs similar to those of Reverend Taylor, and, like him, she has undergone a process of transformation—albeit hers has been much more gradual and less dramatic. An aging widow, a laundress, Aunt Sue lives to support her two sons, Sug and Johnny-Boy—both of whom are committed Communists dedicated to the task of organizing black and white sharecroppers in the region. Sug was arrested and beaten by the local authorities a year earlier and still languishes in jail; Johnny-Boy has continued their clandestine efforts.

The story begins on a stormy night, as Aunt Sue irons while she fretfully awaits the return of Johnny-Boy, who has taken on the assignment of notifying his comrades about a party meeting the next day. While she waits, fragments of religious songs she learned as a child spill from her lips:

> Guiltily, she stopped and smiled. Looks like Ah jus cant seem t fergit them ol songs, no mattah how hard Ah tries. . . . She had learned them when she was a little girl living and working on a farm. Every Monday morning from the corn and the cotton fields the slow strains had floated from her mother's lips, lonely and haunting; and later, as the years had filled with gall, she had learned their deep meaning. Long hours of scrubbing floors for a few cents a day had taught her who Jesus was, what a great boon it was to cling to Him, to be like Him and suffer without a mumbling word. She had poured the yearning of her life into the songs, feeling buoyed with a faith beyond this world. The figure of the Man nailed in agony to the Cross, His burial in a cold grave, His transfigured Resurrection. . . .[23]

This vision of Christian suffering and martyrdom has sustained her through much of her life, particularly in her encounters with the "cold white mountain" represented by the white world and its laws—a test of faith that she has successfully met: "she had grown to love hardship with a bitter pride; she had obeyed the laws of the white folks with a soft smile of secret knowing." It has allowed her to stoically accept, too, the deaths of her mother and her husband, experiences "she bore with the strength shed by the grace of her vision."[24]

Under the tutelage of her sons, however, this Afro-Christian worldview has become systematically dislodged:

> And day by day her sons had ripped from her startled eyes her old vision, and image by image had given her a new one, different, but great and strong enough to fling her into the light of another grace. The wrongs and sufferings of black men had taken the place of Him nailed to the Cross; the meager beginnings of the party had become another Resurrection; and the hate of those who would destroy her new faith had quickened in her a hunger to feel how deeply her new strength went.[25]

A new convert to the party, Aunt Sue is eager to test the strength of her commitments; at the same time, she continues to slip into the deeply entrenched habits and beliefs of her past life, and her deepest instincts remain those of an aging, black woman in the Jim Crow South. Her reveries are interrupted by the arrival of Reva, "a thin, blond-haired white girl," the daughter of one of Johnny-Boy's comrades, with a warning that her father's house—where the meeting is scheduled—is being watched by the sheriff and his men; all the comrades must be notified that the meeting has been canceled. Johnny-Boy will have to warn them.

Aunt Sue's instincts tell her that there is an informant in the party—and it must be a white man. When Johnny-Boy returns, she resumes a long-standing argument between them:

> "Yuh know, Johnny-Boy, yuh been takin in a lotta them white folks lately. . . ."
> . . . . "Ah know whut yuh gonna say, ma. N yuh wrong. Yuh cant judge folks jus by how yuh feel bout em n by how long yuh done knowed em. Ef we start that we wouldn't have *nobody* in the party. . . ."
> "But ain nona our folks tol, Johnny-Boy. . . ."
> She saw his hand sweep in a swift arc of disgust.
> "*Our* folks! Ma, who in Gawds name is *our* folks?"
> The folks we wuz born n raised wid, son. The folks we *know.*"
> . . . ."Ma, when yuh start doubting folks in the party, then there ain no end."
> "Son, Ah knows ever black man n woman in this parta county," she said. . . . "Ah watched em grow up; Ah even heped birth n nurse some of em; Ah knows em *all* from way back. There ain none of em that *coulda* tol! The folks Ah know jus don open they dos n ast death t walk in!"
> . . . . He shook his head and sighed.
> "Ma, Ah done tole yuh a hundred times. Ah cant see white n Ah cant see black," he said. "Ah sees rich men an Ah sees po men."[26]

Aunt Sue's hard-won racial wisdom will be validated by the end of the story—it seems clearly superior to the ideological purity that is Johnny-Boy's blind spot and will become the cause of his undoing—but, interestingly, it does not prevent her from making an exception for Reva, who clearly is in love with Johnny-Boy, and who is credited by the narrator with restoring to Aunt Sue the sense of humanity wrested away from her by racial oppression. In fact, the story confers upon Reva a significance that seems far out of proportion to the limited role she performs in the story—almost as if Wright is determined, as Keneth Kinnamon argues, to suggest "the new mood of the southern proletariat,"[27] and, along the way, to insinuate that Communist organizing efforts in the South *will* have the effect of fulfilling the deepest nightmare of southern racists: increased social intimacy between blacks and whites.

> Reva believed in black folks and not for anything would she falter before her. In Reva's trust and acceptance of her she had found her first feelings of humanity. Reva's love was her refuge from shame and degradation. If in the early days of her life the white mountain had driven her back from the earth, then in her last days Reva's love was drawing her toward it.[28]

Johnny-Boy plunges back into the rainy night to warn his comrades. Meanwhile the sheriff and his rag-tag posse invade Aunt Sue's home, pillaging her kitchen and demanding information about Johnny-Boy and his comrades. Not only does Aunt Sue defy the sheriff and his men, she taunts them and is brutally beaten as a result. Right before he beats her into unconsciousness, the sheriff threatens her: "ef we hafta find im, we'll kill him! Ef we hafta find him, then yuh git a sheet t put over im in the mawnin, see? Git yuh a sheet, cause hes gonna be dead!"[29]

Aunt Sue struggles back into consciousness only to confront the terrifying specter of a "vast white blur" hovering over her, a "huge white face that slowly filled her vision," the face of Booker, a recent recruit to the party. Booker informs her that Johnny-Boy has been apprehended and is being held and interrogated, that is, tortured, in Foley's Woods. The other comrades have not been warned and Booker persuades Aunt Sue to reveal their names to him. Right after Booker leaves, Reva arrives with the devastating news that he is an informer. Revealing nothing to the exhausted young woman, Aunt Sue persuades Reva to go to bed and then devises a plan, fueled by the tension between the two poles of her faith:

> She stood up and looked at the floor while call and counter-call, loyalty and counter-loyalty struggled in her soul. Mired she was between two abandoned worlds, living, but dying without the strength that either gave. The clearer she felt it the fuller did something well up from the depths of her for release; the more urgent did she feel the need to fling into her black

sky another star, another hope, one more terrible vision to give her the strength to live and act.[30]

Out of the endless dialectic between these two contending faiths emerges a new synthesis:

> Her whole being leaped with will; the long years of her life bent toward a moment of focus, a point. . . . She stood straight and smiled grimly; she had in her heart the whole meaning of her life; her entire personality was poised on the brink of a total act.[31]

Carrying a winding sheet that conceals Johnny-Boy's gun, she outraces Booker to Foley's Woods, maneuvers herself to the site where her son is being brutally tortured by the sheriff and his posse, witnesses his suffering in silence, and shoots Booker to death as soon as he arrives. Shot immediately afterwards by one of the posse, Aunt Sue faces her impending death with a sense of equanimity, achieving the martyrdom promised by both Christian and Marxist eschatology:

> Focused and pointed she was, buried in the depths of her star, swallowed in its peace and strength; and not feeling her flesh growing cold, cold as the rain that fell from the invisible sky upon the doomed living and the dead that never dies.[32]

When Aunt Sue first arrived at Foley's Woods, she was greeted with derisive sneers from the whites gathered there:

> Yuh thinka lot of that black Red, don yuh?"
> "Hes mah son."
> "Why don yuh teach im some sense?"
> "Hes mah son," she said again.
> "Lissen, ol nigger woman, yuh stand there wid yo hair white. Yuh got bettah sense than t believe tha niggers kin make a revolution. . . ."
> "A black republic," said the other one, laughing.[33]

The familiarity of these characters with key features of Party doctrine at the time—particularly the slogan "Self-Determination in the Black Belt," with its sometimes attendant calls for the establishment of a black republic somewhere in the South that characterized the Party's outlook during its "Third Period"—adds a significant topical dimension to the story and heightens its realism. Contemporary readers of these stories, and not just those on the Left, would have been familiar with these references, too. Critics who have questioned the plausibility of the action in "Fire and Cloud" and "Bright and Morning Star" cannot be familiar with the extensive work of Communist Party organizers during the early 1930s—not only its broad efforts in Party District 16 (North Carolina, South Carolina, and Virginia) and

Party District 17 (Tennessee, Alabama, and Georgia); but its concentrated activities in Birmingham, Alabama; its campaigns around the "Atlanta Six" and Angelo Herndon in Georgia; its work with black and white sharecroppers in Tallapoosa, Alabama in late 1932 that led to a shootout with white authorities and the deaths of several sharecroppers; and, of course, the Scottsboro campaign.[34] Wright tapped deeply into this vein of recent history to create the stories that marked his debut as a writer, laying the groundwork for his controversial masterpiece *Native Son*.

In an important sense, the novel that established Richard Wright as the first best-selling black novelist in twentieth-century American literary history resumed his interrogation of the relationship between blacks and Communists that he had initiated in "Fire and Cloud" and "Bright and Morning Star."

Sullen and abusive towards his family and friends, subject to sudden and violent mood swings, suspicious and fearful of whites, Bigger Thomas is a product of American society, yet isolated from its mainstream—even as his subjective consciousness is shaped by the same forces that shape the lives of the majority. A twenty-year-old Mississippi migrant to Chicago, Bigger lives in the confines of a racially segregated urban environment, but his deepest instincts remain those of a refugee from the norms of southern Jim Crow life; indeed, Wright's indictment of American life through the pages of *Native Son* rests in his belief that, finally, the circumstances African Americans face in the urban ghettoes of the North are not fundamentally different from the conditions they fled in the South.

Badgered by his mother, who in turn has been pressured by the relief agency, Bigger reluctantly accepts a job as a chauffeur for the Daltons, a wealthy South Shore family known for its philanthropy. Mr. Dalton, we subsequently learn, has made his fortune as a slumlord in Chicago's black community; Mrs. Dalton is blind; and their daughter, Mary, a rebel against her family's bourgeois and liberal values, imagines herself to be a Communist. On the first night of his job, Mary persuades Bigger to pick up her boyfriend, Jan Erlone, a dedicated Communist, and accompany them on an excursion to the black community. To Bigger's acute embarrassment, they insist that he join them at the restaurant he suggests, Ernie's Kitchen Shack, where they exacerbate his growing tension by sharing food and drink with him. The awkward social intimacy they have established continues on the ride back home, as Mary and Jan, now nestled in the back seat of the car, pass a bottle of rum to Bigger to swig as he watches them make out through the rear view mirror. Bigger drops Jan off at a streetcar station and Mary, now very drunk, sprawls next to him in the front seat. She is too drunk to stand, and Bigger, apprehensive, half-drags her into the house and, finally, carries her bodily into her bedroom. In the decisive scene of the novel:

> He lifted her and laid her on the bed. Something urged him to leave at once, but he leaned over her, excited, looking at her face in the dim light,

not wanting to take his hands from her breasts. She tossed and mumbled sleepily. He tightened his fingers on her breasts, kissing her again, feeling her move toward him. He was aware only of her body now; his lips trembled. . . .[35]

The bedroom door suddenly creaks open and Mrs. Dalton—a white blur—appears.

Fearing that if Mary speaks, Mrs. Dalton will discover his presence, Bigger stifles her drunken mumbles with a pillow and kills her. In the immediate aftermath of this event, Bigger enlists his girlfriend Bessie in an ill-conceived plan to extort money from the Daltons; lingers at the Dalton house long enough to be exposed as Mary's killer; flees with Bessie, whom he rapes and murders; and becomes the subject of a city-wide manhunt involving eight thousand pursuers before he is dramatically captured. Although readers of *Native Son* are positioned to know that Mary's death is an accident, Bigger comes to claim it as his own; everything else that happens in *Native Son* is an exploration of the social, political, and ethical implications of Bigger's act.

We know now, since the publication of the restored edition of *Native Son* by the Library of America in 1991, that the bedroom scene was initially depicted in much more graphic terms than those encountered by readers of the Book-of-the-Month Club–sanctioned version of the novel—one of a number of editorial changes that had the accumulated effect of making Bigger Thomas appear to be much more of a potential black rapist than his circumstances in this scene suggest.[36] Such was the price, apparently, for best-seller status. Even the changes proposed by the Book-of-the-Month Club, however, could not completely diminish the force of Wright's intentions to breach the widespread social and cultural taboos against sexual contact between black men and white women. As in "Between the World and Me" and "Big Boy Leaves Home," the old equation of white women = death for black men was central to the drama—and the widespread appeal—of *Native Son*.

Confirming the truth of Wright's convictions, halfway through the composition of *Native Son*, the Robert Nixon case occurred in Chicago.[37] On May 27, 1938, the eighteen-year-old black man was caught by a white woman, Mrs. Florence Johnson, breaking into her apartment. He beat her to death with a brick. Apprehended shortly after and identified by his fingerprints and bloodstained clothes, Nixon implicated another black youth, Earl Hicks. The Chicago police apparently coerced Nixon into confessing to other crimes, too—sexual assaults on two white women and five attempted murders. The *Chicago Tribune* and other newspapers immediately transformed the murder of Mrs. Johnson into a sexual crime—placing the issue of rape at the forefront of its reportage—and singled out Nixon for particularly virulent coverage, routinely referring to him by such terms as "ape-like," "Negro rapist," "rapist slayer," and "jungle beast."

From his new base in New York City, Wright asked his friend Margaret Walker to maintain and send him a clipping file about the investigation and trial of Nixon and Hicks—testimony that he subsequently incorporated word for word into *Native Son*. The International Labor Defense quickly stepped in to assist the black lawyers provided by the National Negro Congress, appointing white attorney Joseph Roth to aid in the defense of Nixon and Hicks. The Attorney General who prosecuted the case, Thomas J. Courtney, provided Wright with the model for Buckley in *Native Son*. The lurid press accounts made the Nixon-Hicks case a local *cause célèbre*, inflaming public passions to the extent that the venue of the trial had to be changed.

Like the Scottsboro case, the Nixon-Hicks trial occurred against the backdrop of confessions and dramatic recantations, accusations about police brutality, and threats of public violence—including a savage physical attack on Nixon at the inquest by the bereaved husband of the victim, Elmer Johnson. In spite of some discrepancies in the evidence, the trial was finished in slightly more than a week, and the jury returned a verdict of guilty after one hour of deliberation. Nixon was sentenced to death. The National Negro Congress quickly sprang into action to mobilize public support for an appeal, enlisting the aid of prominent black Chicago ministers. They won ten reprieves for Nixon before he was electrocuted in the Cook County jail on June 16, 1939. For Wright, there could be no better example of how closely real life mirrored fiction, and vice versa. The Nixon-Hicks trial, the Scottsboro case, and Wright's recurring nightmares all converged in *Native Son*.

If one key to Bigger Thomas's personality is his panic-stricken reaction to the threat of being discovered in a white woman's bedroom, another is his ambivalent relationship to Communism and the Communists who play such a central role in his journey towards self-discovery. Although Bigger is denied access to the mainstream of American life, he participates in some of its values, particularly those circulated within American popular culture through newspapers, advertisements, radio, and the movies. He is subject to the same prejudices and stereotypes about Communists as many of his fellow citizens. In a section of the novel omitted from the original 1940 and all later editions, Bigger's decision to take the job at the Daltons' is cemented when he watches a newsreel featuring Mary Dalton and her radical boyfriend at the movie theater he goes to with Jack. Their ensuing conversation is a mélange of political misinformation and sexual speculation:

Say, Jack?"
  "Yeah."
  "What's a Communist?"
  "Damn if I know. It's a race of people who live in Russia, ain't it?"
  "That guy who was kissing old man Dalton's daughter was a Communist and her folks didn't like it."
  "Rich people don't like Communists."
  "She was a hot-looking number, all right."

"Sure," said Jack. "When you start working there you gotta learn to stand in with her. . . .

"Shucks, my ma use to work for rich white folks and you ought to hear the tales she used to tell. . . ."

"What kind of tales?" Bigger asked eagerly.

"Ah, them rich white women'll go to bed with anybody, from a poodle on up. They even have their chauffeurs. . . ."[38]

This scene prefigures—and predetermines—Mary's death and Bigger's fate. Moreover, it establishes Communism as an integral feature of Bigger's environment—even if for Bigger the Communist Party initially is simply a means of diverting suspicion from himself to an organization he shrewdly recognizes is at least as despised as he is. While Bigger is not a Communist—and, in fact, resists their attempts to define his being through the last pages of the novel—Richard Wright most decidedly *is*, and his knowledge of the Party, particularly its official perspective towards African Americans, is central to the plot and themes of *Native Son*.

The name "Mary Dalton" is an inside joke, for example, Wright's deliberate appropriation of the name of one of the white defendants of the "Atlanta Six," a group of Communists arrested for distributing anti-lynching leaflets in Georgia. Held in jail in 1930–31, they faced the prospect of being electrocuted for violating the Georgia Insurrection Act if they were convicted. The Communist Party spearheaded a national campaign that eventually led to their release on bail. After her release, Mary Dalton was assigned by the Party to work in Chicago, where, according to Michel Fabre, Wright took a hearty dislike to her: "it gave him pleasure to think that people like Harry Heywood and John Davis"—both prominent African American Party leaders—"would get the subtle allusion if he used her name in this context."[39]

To compound the irony, Mary Dalton had been closely associated with Herb Newton, a prominent black Communist on Chicago's South Side—and the only black member of the "Atlanta Six." Herb Newton's white wife, Jane—an exact contemporary of Wright's and a close, platonic friend—had been raised in an upper-middle-class, Republican family in Grand Rapids, Michigan. A rebel against the privileges of her social class, Jane had dropped out of college, married, moved to New York, joined the Communist Party, divorced and remarried, and divorced again before marrying Herb Newton. They were famous in Chicago because of their participation in a 1934 eviction action that led to their arrests. Recognized in court by an attorney as a member of a "good family," Jane was sent to a psychiatric hospital for ten days' observation by a judge who clearly questioned the sanity of a white woman of her background who was a committed Communist and had married an African American.[40] After a stint of several months in the Soviet Union in 1937, Jane and her husband moved to Brooklyn. After he settled in New

York, Wright lived with the Newtons from May to October 1938, where he worked furiously on the composition of *Native Son*—with Jane looking over his shoulder as reader and critic; they sometimes had heated exchanges about the relative weight of Wright's aesthetic and political choices.

In this context, Wright's characterizations of the two key Communists—Jan Erlone and the Jewish attorney and Labor Defender representative Boris Max—are critical to his design. Of the two, Jan—modeled after one of Wright's friends, Jan Wittenber, a member of the Chicago John Reed Club and secretary of the Illinois State International Labor Defense—is the more noble and selfless.[41] In the critical third section of *Native Son*, "Fate," an extended meditation on Bigger Thomas's capacity to be "saved"—in the sense of gaining deeper insight into his relationship to other human beings—Jan represents the first breach in the "looming mountain of white hate" that has oppressed Bigger throughout his conscious life:

> Jan leaned forward and stared at the floor. "I'm not trying to make up to you, Bigger. I didn't come here to feel sorry for you. I don't suppose you're so much worse off than the rest of us who get tangled up in this world. I'm here because I'm trying to live up to this thing as I see it. And it isn't easy, Bigger. I—I loved that girl you killed. . . . I was in jail grieving for Mary and then I thought of all the black men who've been killed, the black men who had to grieve when their people were snatched from them in slavery and since slavery. I thought that if they could stand it, then I ought to.[42]

Jan's struggle to forgive Bigger, or—at least—to place his actions in a meaningful context, takes on the ring of Christian values, in sharp contrast to the behavior of the black minister Reverend Hammond, whose religious pieties quickly slide into other-worldly slogans. Ironically, in view of Wright's sympathetic portrayal of Reverend Taylor in "Fire and Cloud," the character of Reverend Hammond in this passage reverts to the stock portrayal of black preachers in proletarian literature. If Bigger Thomas is to achieve salvation, clearly it will have to come from sources other than the black church.

Jan's sincerity, unfortunately, can do little to ameliorate Bigger's circumstances—except to introduce him to attorney Max—and his testimony at the inquest conducted by the Deputy Coroner does nothing except inflame the passions of the people gathered there.

The second, and more important, crack in the "white mountain" occurs in the relationship that develops between Bigger and Max—or "Mr. Max," as Bigger calls him—a character that some critics argue is based upon the figure of Samuel S. Leibowitz.[43] Wright emphasizes that Max's Jewishness confers upon him a degree of compassion and insight that sets him aside from many whites, but Max's primary function is as a spokesman for the Communist Party's point of view—notable in the first instance for its willingness to see Bigger's apparently individual plight in terms of a larger nexus of social

and political relationships. Never missing an opportunity to offer Bigger a
political education, Max makes this clear at the very beginning of their extended conversations:

"They hate black folks," he [Bigger] said.
"Why, Bigger?"
"I don't know, Mr. Max."
"Bigger, don't you know they hate others, too?"
"Who they hate?"
"They hate trade unions. They hate folks who try to organize. They hate
Jan."
"But they hate black folks more than they hate unions," Bigger said.
"They don't treat union folks like they do me."
"Oh, yes they do. You think that because your color makes it easy for
them to point you, segregate you, exploit you. But they do that to others,
too. They hate me because I'm trying to help you. They're writing me let-
ters, calling me a 'dirty Jew.'"
"All I know is that they hate me," Bigger said grimly.[44]

Max's kindly but paternalistic attitude towards Bigger prefigures the rhetoric he will adopt in his lengthy, eloquent, and passionate defense, where he
acknowledges Bigger's guilt but insists upon portraying him as an untu-
tored—by dint of his systematic exclusion from it—product of American life.
Ironically, Max's persistent allusions to Bigger as a "boy," a "Negro boy," a
"black boy," and a "poor black boy" dimly echo the rhetoric surrounding the
Scottsboro case, prompting Buckley to protest: "I object to the counsel for the
defendant speaking of the defendant before this Court by any name other
than that written in the indictment. Such names as 'Bigger' and 'this poor
boy' are used to arouse sympathy."[45] Still, Max's defense of Bigger is not without its flaws—some of them fatal, as shrewd readers of *Native Son*, like black
Party official Benjamin Davis, Jr., were quick to point out.[46]
For readers who viewed the defense of the Scottsboro boys as a textbook
case for how an effective mass defense should be mounted, Max's decision to
plead Bigger guilty, forego a trial by jury, and—in effect—throw his case on
the mercy of the Court is a blatant violation of the sacred principles of the
International Labor Defense. Like the Scottsboro boys, Bigger Thomas is a
character of limited social and political consciousness who is not engaged in
class struggle, has little knowledge of it, and does not identify with it. All the
more reason then to orchestrate a full-court mass-defense on his behalf. The
cornerstone of the ILD defense strategy is legal action inextricably linked to
a mass campaign. District Attorney Buckley effectively incorporates aroused
popular sentiment into his argument—even when he is chastised for doing
so by the Court; Max does not—as if he has concluded in advance that it either does not exist or, if mobilized, will not have the force to overcome the

apparent will of the majority. Neither does he make even a token gesture towards challenging the composition of the jury—a precedent clearly established by the second Supreme Court decision to emerge out of the Scottsboro case: *Norris v. Alabama*, the 1935 decision that overturned the Scottsboro convictions because African Americans had been excluded from Alabama juries.

Moreover, the defenders of the Scottsboro boys recognized—or believed—that an effective mass defense could not be launched without taking direct aim at the racist mythology that fueled the terror unleashed upon black men: the specter of the black rapist. Scottsboro defenders had addressed this issue, in part—as we have seen—by insinuations, or bold attacks, on the class background and character of the accusers, Ruby Bates and Victoria Price. Wright made the issue infinitely more complicated for Max not only by creating a victim from the upper-middle class, but also by explicitly acknowledging Bigger's sexual attraction to his victim—and by leaving his defense exclusively in the hands of one man. For Max, the sexual issues pale in the face of the deeper significance of Bigger's existence: "Let us not concern ourselves with the part of Bigger Thomas's confession that says he murdered accidentally, that he did not rape the girl. It really does not matter. . . ."[47]

In the first three stories of *Uncle Tom's Children*, Wright implicitly criticizes the individualism of his characters, dramatically underscoring their helplessness against social and political forces that finally overwhelm them; in *Native Son*, Max's legal strategy seems premised on the belief that he alone has the capacity to save Bigger's life.

In this regard, Max's rhetorical posture is a curious one. Speaking in the first person, he implicitly aligns himself with the "civilized" values of the Court, thereby widening the gap between Bigger and society, rather than narrowing it.

Rather than arguing for Bigger's life on the basis of the evidence, Max proposes, in effect, to provide a deeper critical exegesis, a kind of "social psychotherapy" that argues America itself is a patient badly in need of psychiatric care:

> "I say, Your Honor, that the mere act of understanding Bigger Thomas will be a thawing out of icebound impulses, a dragging of the sprawling forms of dread out of the night of fear into the light of reason, an unveiling of the unconscious ritual of death in which we, like sleep-walkers, have participated so dreamlike and thoughtlessly."[48]

Although this section of *Native Son* has often been singled out by critics as the most blatant example of how the intrusion of Communist ideology has marred Wright's novel, Max's speech is fascinating for its shifts and turns—*and* for its deviations from the official Party line. Arguing that Bigger represents "a mode of *life* in our midst . . . human life draped in a form and guise alien to ours, but springing from a soil plowed and sown by our hands,"[49] Max dusts off the "Black Nation" thesis—"Taken collectively, they are not simply twelve million people; in reality they constitute a separate nation, stunted, stripped, and held captive *within* this nation, devoid of political, social, economic, and property rights"[50]—even though the Party had quietly discarded its "Self-Determination in the Black Belt" slogan after it entered its "Popular Front" phase in the mid-1930s. Rising to new heights of rhetoric, Max proclaims: "They glide through our complex civilization like wailing ghosts; they spin like fiery planets lost from their orbits; they wither and die like trees ripped from native soil."[51] Under these circumstances, the fate of people like Bigger is preordained: *"His very existence is a crime against the state!"*[52]

Having depicted Bigger in such chilling and prophetic terms, Max compounds his political apostasy by arguing that characters such as he could as easily be drawn to Fascism as to Socialism—thereby challenging the fixed Party belief in the revolutionary potential of the black masses; and he further complicates his argument by insisting that Bigger is sometimes white, too: "There are others, Your Honor, millions of others, Negro and white, and that is what makes our future seem a looming image of violence."[53]

Max's solution to this apocalyptic vision? Life imprisonment: "You would be for the first time conferring *life* upon him. He would be brought for the first time within the orbit of our civilization. He would have an identity, even though it be but a number. He would have for the first time an openly designated relationship with the world. The very building in which he would spend the rest of his natural life would be the first recognition of his personality he has ever had."[54]

Needless to say, Max's rhetoric does not sway the Court—and Bigger does not understand it, even though he appreciates Max's efforts. In fact, until the last paragraph of *Native Son*, Bigger steadfastly resists the role conferred upon him in Max's interpretive universe, gaining in the end a measure of self-affirmation:

> "When a man kills, it's for something. . . . I didn't know I was really alive in this world until I felt things hard enough to kill for 'em. . . . It's the truth, Mr. Max, I can say it now, 'cause I'm going to die. I know what I'm saying real good and I know how it sounds. But I'm all right. I'm all right when I look at it that way."[55]

In a novel dominated by images of blindness, Max's final gesture effectively captures the gulf between himself and Bigger:

> Max groped for his hat like a blind man; he found it and jammed it on his head. He felt for the door, keeping his face averted. He poked his arm through and signaled for the guard. When he was let out he stood for a moment, his back to the steel door.[56]

Max is a nominal Communist, but, in the final analysis, he is less a spokesman for the official point of view of the Communist Party than he is a figure through whom Wright channels his own perspective. The final exchange between Max and Bigger, however, does signify the ways in which the Communist Party, in Wright's view, continued to miss the mark in its attempt to court African Americans. In *Native Son*, Wright seizes upon the classic "rape/lynching" narrative not only to offer a damning critique of "the Negro's uncertain place in America," but also—as he had in *Uncle Tom's Children*—to continue to insist that the "Negro Question" still cried out for solutions.

Wright's obsession with the peculiar convergence of race, sexuality, and racism that was at the root of the Scottsboro case would continue to fuel his imagination even after he departed the United States for France in 1947, reappearing in the last novel he published during his lifetime, *The Long Dream*, and in his posthumously published story "The Man Who Killed a Shadow."[57] In the hands of the generation of African American writers that emerged during the post–World War II era, many of them intent upon pursuing "raceless" and "universal" themes in their writing (in itself a response to growing Cold War anxieties[58]), however, Scottsboro became a fading reference. William Gardner Smith's 1948 novel *Last of the Conquerors* tells the story of a black U.S. soldier in the army of occupation in Germany, who discovers that there are greater opportunities for racial democracy in post-Hitler Germany than there are in the United States. In one critical conversation, as several black soldiers mull over their options after their terms of enlistment end, one of them comments:

> "I guess I can tell you guys. I ain't going back to the States, pals. I won't be here when the shipment leaves."
>
> "What do you mean? Where will you be?"
>
> "I got it all figured out. See, about two months ago I was in the Russian sector and this Russian officer approached me at the bar and brought me a drink. We sat down and started talking about things in general and after a while we got around to the race question in the States. This officer knew more than I did about it. He could name names and call dates, like the Scottsboro boys case and the Philly PTC strike. He asked me why I didn't go to Russia to live. I told him no, I wasn't for that. He said if I ever wanted anything I could come to him and he gave me the name of his company. So now I figure I might as well go into the Russian sector of Berlin to live for a while."
>
> "You mean desert?" Randy asked unbelievingly.
>
> "That's right. I'm gonna live in the Russian sector. Can't no American MP's come in there and get me."
>
> "But you won't never be able to return to the States," Randy said.
>
> "I know it."[59]

If William Gardner Smith treats Scottsboro as part and parcel of the appeal of Soviet Union propagandists to disillusioned black American GI's, William

Demby's 1950 novel *Bettlecreek* invokes the Scottsboro case to underscore the bleak, existential terms of his fictional universe. Set in the small town of Bettlecreek, West Virginia, the novel places at its center a tragic white victim named Bill Trapp, a retired carnival worker, a social recluse who lives on a ramshackle farm between the black and white parts of town. Around him revolve two black characters: fourteen-year-old Johnny Johnson, who has been forced by his mother's illness to join his uncle and aunt in Bettlecreek; and David Diggs, Johnny's uncle, a frustrated artist trapped by the provinciality and spiritual deadness of his community. Out of his profound sense of loneliness, Bill Trapp reaches out to Johnny and David respectively, and when they respond to him, he decides to continue to extend his humanity, first by donating pumpkins to an African American church festival and, more importantly, by organizing a racially mixed picnic for children in the Bettlecreek community. The picnic is an unmitigated disaster, and, in its aftermath, malicious rumors begin to circulate in the town that Trapp is a child molester. In the meantime, David Diggs moves towards the fateful decision to abandon his wife and leave town with his childhood sweetheart, who has returned to Bettlecreek for the funeral of her guardian; while Johnny, lonely, acutely aware of his tenuous position as a newcomer to the community, and eager for social acceptance, falls under the spell of a community gang called, ominously enough, the Nightriders—complete with a brutal leader, black robes, and excruciating initiation rituals. As the mass hysteria surrounding the false accusations of sexual molestation begins to reach a fever pitch, a critical conversation occurs in Bettlecreek's black barbershop:

> Slim said indignantly, "Now you turn the tables of this thing. Make like there was a colored man lived all by hisself in a white part of town and he gets it in his noodle to bring the races together just like he's Jesus Christ, and I don't believe that was the reason to have the picnic in the first place, regardless of what that Diggs is going around sayin, but just play like he did want to bring the races together and it was a colored man livin in a white town, and he starts all this monkey business with white girls mind ya, what do you think would happen to *him*? You just pick up any newspaper from the last few years and read what happened to them Scottsboro boys and you can imagine what'd happen to him! His life wouldn't be worth a row of pins I mean to tell ya—not one row of pins!"[60]

Slim's invocation of Scottsboro as a rallying cry is one of a chain of events that leads to the tragic denouement of the novel—Bill Trapp's death after the Nightriders command Johnny to burn down his house as the price of his initiation into the gang.

David Diggs's moral cowardice, his refusal to contest the insidious barbershop gossip when he is clearly in a position to dispute public opinion, and Johnny's outright betrayal of his growing friendship with Bill Trapp in his eagerness to win acceptance by the Nightriders, constitute two sides of the same coin—a depth of betrayal underscored by the Christian imagery associated with Trapp. The invocation of Scottsboro in the midst of their respective moral evasions displaces the historical meanings of the case onto the shoulders of the feckless white victim, Bill Trapp, suggesting in the process how easily the ritual of scapegoating can occur in radically different social and historical circumstances—thus "universalizing" the experience. From Demby's determinedly existential viewpoint, Trapp's fate—the easy reversal of racial roles in this particular tragic drama—is symptomatic of the larger issue of human isolation and alienation in a world devoid of meaning.

Although William Gardner Smith and William Demby followed in Wright's footsteps, seeking their lives and careers as expatriates in Europe during the 1950s, they left the Scottsboro case behind when they departed the United States.

CHAPTER SIX

THE SCOTTSBORO DEFENDANT
AS PROTO-REVOLUTIONARY

Haywood Patterson

> I always protested when I didn't like things. Down South I always talked
> like I wanted: before Scottsboro, during it, and since.
>
> —Haywood Patterson, *Scottsboro Boy*[1]

> Patterson and I didn't really get along too good because every time you
> turn around he'd be into something or another with the rest of us. He
> just wasn't liked by most of us. . . . I didn't like his ways and I didn't
> like him. He was the kind of person you couldn't tell nothing to. He
> was overbearing, always ready to fight, always into something.
>
> —Kwando Mbiassi Kinshasa, *The Man from Scottsboro: Clarence
> Norris in His Own Words*[2]

> Haywood Patterson . . . was apparently chosen by the prosecution to be
> tried first because he has the blackest skin, the wickedest gleam in his
> eyes, and the meanest expression on his face. He is what is known in
> the South as a "bad nigger." This means that he is willful, self-assertive,
> independent, not properly servile.
>
> —Carleton Beals[3]

IF THERE IS INDEED a dialectical relationship between art and life, there is no
more striking symmetry than that of Scottsboro defendant Haywood Patterson
and Richard Wright's fictional creation Bigger Thomas. In the final scene of
Native Son Bigger Thomas, alone, grasps the bars of his prison cell and smiles
"a faint, wry, bitter smile," apparently content in his recognition that he has
achieved a measure of psychological and spiritual freedom in spite of his
physical confinement. For Haywood Patterson, who, like his fictional counterpart, was sentenced to death for a crime he did not commit, such a resolution to his trials would have been unimaginable—too great a capitulation to
the forces that sought to annihilate him.

Defiant, sullen, wily, suspicious, and capable of sudden and violent mood swings, Haywood Patterson was arguably the central figure among the Scottsboro defendants. It was, after all, a sharp exchange between Patterson and a white hobo that triggered off the subsequent altercation on the train heading into Alabama:

> The trouble began when three or four white boys crossed over the oil tanker that four of us colored fellows from Chattanooga were in. One of the white boys, he stepped on my hand and liked to have knocked me off the train. . . .
>
> I made a complaint about it and the white boy talked back—mean, serious, white folks southern talk.
>
> That is how the Scottsboro case began . . . with a white foot on my black hand.
>
> "The next time you want by," I said, "just tell me you want by and I let you by."
>
> "Nigger, I don't ask you when I want by. What you doing on this train anyway?"
>
> "Look, I just tell you the next time you want by you just tell me you want by and let you by."
>
> "Nigger bastard, this is a white man's train. You better get off. All you black bastards better get off."
>
> I felt we had as much business stealing a ride on this train as those white boys hoboing from one place to another looking for work like us. But it happens in the South most poor whites feel they are better than Negroes and a black man has few rights. It was wrong talk from the white fellow and I felt I should sense it into him and his friends we were human beings with rights too.[4]

Patterson's truculence, his uncompromising insistence upon his rights and fair play, earned him the enmity of the Alabama judicial system. Singled out as the first target for prosecution, he was convicted four times between 1931 and 1937 and sentenced to death three times; he also served the hardest time of all of the Scottsboro defendants, including a six-year stretch at the notoriously brutal Atmore State Prison Farm in southern Alabama—"the southmost part of hell."[5] At the same time, Patterson's defiant posture made him one of the most consistent and loyal supporters of the ILD—even as he sometimes railed against the organization in his letters to its representatives—and, therefore, a convenient symbol and rallying point for Scottsboro defenders.

Patterson escaped from Atmore in April 1943, managing to elude his pursuers for five days before he was recaptured. Shipped back to Kilby Prison, Patterson endured the brutal conditions there for five years before he escaped again, drowning two hound dogs in the process. He headed for Atlanta, then

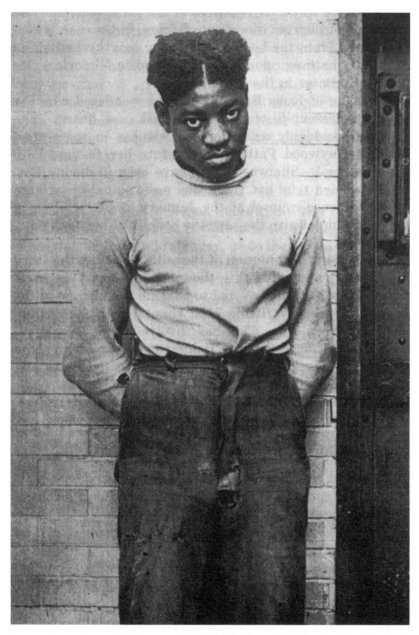

Figure 6.1 Haywood Patterson: "Four juries found him guilty." Source: Files Crenshaw and Kenneth A. Miller, *Scottsboro: The Firebrand of Communism* (Montgomery, AL: Brown Printing Company, 1936).

Chattanooga, and Detroit, where his sister lived. Somehow, word of Patterson's presence in Detroit leaked out and he fled further east.

In the spring of 1949 Patterson had a fateful encounter with the crusading investigative reporter I. F. Stone, then on the staff of the *New York Post* as a Washington columnist. Stone's editor, Ted O. Thackrey, had received "a mysterious phone call from a well-known Negro leader,"[6] most likely Congressman Adam Clayton Powell, Jr., informing him that one of the Scottsboro Boys had escaped from prison and was hiding in Harlem. Anticipating an extradition battle, the caller was seeking a journalist who could do a series of background articles on the case. Assigned by Thackrey to follow up the story, Stone made his way to Harlem, where he had a clandestine meeting with Haywood Patterson and writer Earl Conrad "in a dark apartment, with extra chains on the door and a huge mastiff as guard."[7] Stone described the scene in graphic if somewhat stereotypical terms: "In a dark corner, the white balls of his eyes wary and alert as those of a hunted creature, sat a lanky Negro, deep black and tense, is if he might have to start running at any moment."[8] As he listened to Patterson speak "in a weird Alabama dialect I found difficult at first in understanding,"[9] Stone quickly grasped the dramatic and political potential of the story he heard unfolding. He contacted the chief editor of Doubleday, Ken McCormick, and McCormick acted quickly and decisively, providing the cash advance and contract that made it possible for Patterson to leave Harlem with Conrad and his family to re-create his life's story. The result of their collaboration, *Scottsboro Boy*, introduced a new stage in the Scottsboro Narrative: the only contemporary, full-length, first-person narrative to emerge out of two decades of legal and political struggles.

Stone couched his initial encounter with Patterson in the vocabulary of narratives of American fugitive slaves, noting that Patterson had "managed to make his way north like an escaped slave," and adding: "He might have been a fugitive from a 17th century African slave ship."[10] Stone's initial impressions were quite *a propos*, for the harrowing experiences that Haywood Patterson narrated, and Earl Conrad crafted, in *Scottsboro Boy*, placed Patterson squarely within a tradition in African American narrative literature.[11]

Casting Haywood Patterson as the leading figure in the ordeal of the Scottsboro defendants, *Scottsboro Boy* simultaneously reworks the traditional tropes of the "quest for freedom and literacy"[12] so central to African American autobiography into a sharp critique of southern prisons and prison camps—effectively establishing this work as a seminal text of contemporary American prison literature.[13]

Like many of the Scottsboro Boys, Haywood Patterson was illiterate when he entered prison in 1931, noting somewhat ruefully in his autobiography:

> The little training I had had I never much followed up. Never read no papers, no books, nothing. I did not know how to pronounce things, could

not even say "Alabama" so you could understand me. I had to ask prisoners to tell me, and I would try and repeat it.[14]

He confessed a similar inability to write: "In the death cell [in 1931] I held a pencil in my hand, but I couldn't tap the power that was in it. I couldn't write. I couldn't spell."[15]

Armed with the gift of a Bible from a fellow inmate, however, Patterson resorted to various ruses as he began his trek towards greater mastery of the written word—somewhat reminiscent of Frederick Douglass's accounts of his "theft" of literacy in his *Narrative* and Malcolm X's odyssey towards greater knowledge in his *Autobiography*:

> Now, with the bible reading, I got to see how words were spelled. I spelled them out loud, and once I spelled a word right I never forgot it. From a small dictionary I had I got to know all the capitals of the United States.
>
> A person would come along, a guard or a visitor, or anybody, and I would bet a dollar I knew the capital of every state. Then I would name them—and call off the bet. It made me good not to take that kind of money. So I had two friends, the dictionary and the Bible. When I came on a word where the meaning I didn't know I switched from the Bible to the dictionary.[16]

In his autobiography, Patterson made a direct connection between his pursuit of learning and his refusal to accept the inevitability of his death sentence, asserting with characteristic bravado: "The reason I took such an interest in learning was this. I never believed I was going to die. A condemned man who knows his time is up, he will not take up something new."[17]

By the end of 1931, Patterson had inaugurated his own correspondence, first to his mother and then, during the ensuing years, to a widening circle of national and international figures that included William Patterson, the African American Communist attorney who was appointed the National Secretary of the ILD in 1932; Kay Boyle, the avant-garde American writer then living in France; Nancy Cunard, the flamboyant and rebellious shipping-line heiress; Allan Knight Chalmers and Morris Shapiro of the Scottsboro Defense Committee; and many, many others.

From the very beginning, even as he struggled to master the fundamentals of spelling, grammar, and syntax, Patterson revealed a shrewd sense of audience for his letters, carefully crafting them according to his perception of his reader. Thus, in an early letter to J. Louis Engdahl, then the National Secretary of the ILD, at the height of the early struggle between the ILD and the NAACP over who could best represent the Scottsboro Boys, he seems to be weighing the merits of the Scottsboro defendants being represented by Clarence Darrow:

Dear Sir:

While sitting all alone in prison I thought I'll express you a few lines to let you here from us boys. We all are well and hoping to be free soon and also hoping you all will remain in fighting for us boys.

Mr. Engdahl I am ask you a question and I would like for you to answer it in your write, and here it's are. Have you all got Mr. Darrow to fighting for us boys. The reasons why I ask you that becost I heard that Mr. Clance Darrow was going to be fighting for us boys, and I would like to know if possible becost I am innocent, as innocent as the tiny mite of life just beginning to stir beneath my heart. Honest, Mr. Engdahl, I haven't did anything to be imprisonment like this. And all of the boys send their best regards to you all and best wishes. So I would appreciate an interview at your earliest convenience.

Very truly yours,
Haywood Patterson[18]

Patterson's question was rendered moot, of course, by Darrow's withdrawal from the case in late December 1931. From that point on, Patterson cast his lot with the ILD—as he made plain in a letter to William Patterson:

I only wish I was more educated so that I could express my sincere appreciation to the people in New York for all that they have did for we boys while here in prison. And we are well please in the way you all taking of things for us no matter what they says about the ILD. I have always wish for the ILD to take care of me and allways will so.[19]

Patterson apparently followed published accounts about the Scottsboro case with keen interest, and wrote enthusiastically to playwright John Wexley two weeks after the opening of *They Shall Not Die*:

My dear Wexley Sir

This certainly is a great moment for me to take the pleasure in trying to Write a greater friend as you are of course I hardly realize just how to express my sincere appreciation to you and the congregation for the great trouble which you and all of you are putting forth in my Behalf and I am sure grateful To hear of the play oh I trust that it will be of great success—and it only Gives me great joy To realize that I have the Heartfelt sympathy of you all I hope you know Mr. Liebowitz I love him more than life its self He is such a mighty great man I think such a lot of him in case I should live 19 years from today I will remember him. Now Mr Wexley you will excuse my expression which it are a little Different from yours now I wish the play will Bring forth good results. Please extend my sincere appreciation all of the workers.[20]

Within the confines of the Birmingham jail, where the Scottsboro Boys were often confined between trials, Patterson and the other defendants quickly discovered that their notoriety and their correspondence with the outside world, ironically, conferred upon them a higher social status than they had experienced while they were nominally free in Depression America:

> Yes, we lived easy in this jail, we Scottsboro boys did. We had the money and we had everything we wanted: whiskey, knives, clothes. We could have got pistols if we wanted. . . .
>
> In jail we fellows got dressed up for the first time in our lives. There was a jailer, his name was Crofts. Any time we wanted for anything down the line of clothing, or anything else, he would take a couple hours off. He would go off duty and do our shopping for us. Buy anything we wanted.
>
> The feeling was very high about our case between my second and third trials. Hundreds of dollars came in to us. I ate the same food as the warden and the jailers, just the same. The chef, he would cook our food specially. We paid for it, never had to eat the prison food.
>
> Sometimes we laughed about it, the ups and downs. Outside we were hobos; inside the prison we had some respect. By our being in jail we were getting the food there that our families never was able to buy us. It was a twist of the case we were in. We took advantage because we didn't know when we'd go back to the death row.[21]

Patterson sometimes sought to cultivate his supporters by adopting a formal, somewhat obsequious tone—as he did in a letter to Morris Shapiro of the Scottsboro Defense Committee:

> Mr. Dear Mr. Morris Sir your very cheerful and encouraging letter which contained a $2.00 money order has been gladly received and highly appreciated. Of course at the present it is rather hard on me to express my inmost appreciation and sincere gratitude to you and the workers for making better our condition. While here in this jail it is mighty nice of you workers to add on our monthly allowance which it certainly will be a great Deal of assistence to we boys now that we can have a lot more comforts.[22]

But Patterson's opportunism and, ironically, sense of entitlement sometimes led him to adopt other tactics, as he did in a letter he wrote to Miss Juanita E. Jackson, a representative of the national staff of the NAACP, after she visited the Scottsboro boys with several members of the Birmingham NAACP on November 20, 1936.

Patterson initially adopted a courtly attitude towards Miss Patterson:

> My Dear Juanita E. Jackson:
>
> I suppose you are going to be a little surprised on receiving these lines yet I hope your surprise will be of an agreeable kind. You see I am being true to my promise. I promised to write you but I have just gotten around

Figure 6.2 Sheriff and guardsmen taking Haywood Patterson to Decatur courthouse. Source: Arthur Garfield Hayes, *Trial by Prejudice* (New York: Civici Friede Publishers, 1933).

The Defendant as Revolutionary • 177

to perform it. You know I am a poor writer but none the less I certainly have a great deal of it to do each day of my life daily. . . .

You know to-day I obtained the pittsburgh Courier, a colored newspaper and in it I saw the pictures of us which we had posed with you all. And I am exceedingly glad to see How delightful and beautiful we appear to be in it. You see when you were down to enterview we boys I must admit that I were so excited that I hardly knew what I were doing. Again I acknowledge to the fact that I were before you came in a High State of excitement. Therefore if I acted very silly in the past will you forgive me, and try to know that it wasn't intentional. You see Dear Juanita it is infrequently that we boys have visitors. Hardly ever we do. . . .[23]

He continues in this vein for another page or so before turning to the issue at hand:

Now Juanita I want to suggest that I didn't so very well appreciated your acting when you were down. And that is you seems to show a particular interest in [Olen] Montgomery. what I am referring to is that you seemed to have listen to him and usable him a little money. And gave me nothing for my needs you see I later learned that you gave him ($5.00) in particularly. a heck of a fine sport you are. anyway I wish to suggest that Christmas is nearly at hand and I wonder miserably could you find it in your heart to do me the selfsame courtesy. You see that I would like you to do me a favor and that is send me a little money.[24]

But Haywood Patterson was also capable of writing letters that were eloquent in their expression of his deepest feelings. Such was the case when he learned of the death of his mother, Janie, ten months after the death of his father, Claude:

My Dear Mr. Shapiro

You can well imagine how terribly I was shocked upon receiving the sad news of the Death of my beloved Mother. Now that everything in life to me is gone, the future seems so empty. It holds nothing of useful or great value to me. The one thing only hurt most is that I were not free to ask them to forgive for all the miserable Heartches and sorrows and Heavy Burdens I caused them to bear. . . . you know my father and Mother's life was made miserable for them. And they suffer under a great strain. It certainly is hard for me ever to regain my sanity and balance of power.[25]

Among his numerous correspondents, Patterson seemed to take the most pleasure—and was the least restrained—in his exchanges with Anna Damon, ultimately the National Secretary of the ILD, from the mid-1930s until her death in 1943.

Figure 6.3 Scottsboro Boys and Miss Juanita E. Jackson. Source: Library of Congress.

Patterson wrote freely to Damon about his personal grievances:

I am ashamed for this long silence But the truth is that I have felt so un-
happy and horribly disappointed and resentful over the outcome of my last
tiral and the way the case are now being conducted. Altho I am most certainly that you have felt in my silence too. Now I wrote you a letter some
time ago telling you in particularly about how you your self had taken to
done But you shunned me for a while. I thought nothing of it though and
I need not to have worried for my long silence and delay an I realized that
it came about largely on account of your conduct. Now I wont spare myself
the trouble to comment an I feel that you are not the one to blame. I cant
believe you have done such things that have been told to me and I think I
am going to like you more better now But you must behave your self and
not to bother me too much by telling people untrue things. . . .[26]

Patterson confided in Damon about his daily concerns, particularly his
sometimes troubled relationships with some of the other Scottsboro Boys:

Anna I want to tell you about How things is going on and how some of the
Boys are doing or rather it is only Roy Wright and Charlie Weems. And of
course to tell the truth I have try in every way there is open to me to get
along with Roy Wright especially when I am in his present but that have
seemingly an impossible thing to do and for some reason or other they
have put him in confinement near me in my company and I regret it terribly because he is a most Disagreeable boy. Honestly Roy Wright is an un-
reasonable foolish sort of boy. I never knew he was such a fellow. Why he
is inhuman. And very unkind you see Roy have abuse me terribly on many
occasions. . . . He is a piecebreaker among the boys and myself you see He
always set their hearts against me by saying things concerning me . . . You
see I can not afford to allow him to injure me anymore like an you can recall or remember once about 3 years ago him and his Brother Andy Wright
over power me and cut and beat me tirrible you see the Doctor had to use
six stitches and every since then he have had a extreme dislike. . . .[27]

In another letter, angry that Anna Damon has apparently not sent him a
hoped-for birthday present, Patterson lashes out at the ILD:

You must know that I wish to be of some assistance to you all. I want to be
helpful to you, and utilize the organizations, but it have so seemed that you
prime leaders do not wish to be useful to me in some certain things. and I
am forced to acknowledge that I really loosing confidence in some of you.
I regret to say this But it is best to be truthfully. I saw in Labor Defender
Magazine one of the Horrible looking picture that you all could find to
place in it and says that it was me and the reading above it says send Him
greetings. I don't like it. And I beg of you all to not place another one like

it in the paper or magazine please. . . . I asked Mr. Shapior [Morris Sha-
piro] to stop you all from use slanders and those Horrible pictures against
me Because it is hurtful to my very soul.

In the same letter, Patterson wonders about what has brought him to these
circumstances:

I have thought of myself a great deal and Haven't yet figured out how I am
being unjustly framed and just what can be holding me in prison. Of
course some says for an answer is that C.P. and N.Y. Jews is the bigger reason for my being kept constantly incarcerated and relate that I should keep
a way from them. You will more than liken to be released immediately.
Why I don't know about that you see. I like the Jew race of people and always has, and all my life nearly was spent among them daily and I understand them all too well. Why they have never caused me to grief or to suffer any misfortune. Honestly some of them seems to like me so much an to
try to teach me how to speak some things in their language which is not
lawful. . . . I can never feel against a Jew or even anyone. I like them. They
are really the best sort of people and I gets a great kick out of hearing some
of them talk.[28]

Patterson's sensitivity to real or imagined slights spilled over in a letter he
wrote to Anna Damon in January 1937:

My dear Miss Damon yours was gladly received some few days ago in
which I found enclosed the regular monthly allowance and for which I
thank you, now that I have read this letter with interest.

However I noticed the signature and it have seemed that you did not really written this letter because it wasn't your signature or so it seems to me.
And I don't like that. You are my little pal and I desire having you to write
me always. So stop putting some one else up to do it or I am going to cease
my communication get that?[29]

This outburst was followed an extended account of his depression, apparently triggered by a neuropsychiatric examination—ironically requested by
the Scottsboro Defense Committee, under the leadership of the Reverend
Allan Chalmers—that had been administered the day before:

The doctor came to Enterview we boys yesterday and he gave examination.
Of course he did not disclose anything to me and if there anything Wrong
I don't Know About it. Of course I have thought and worried so much that
my mind is feverish and sick. And I am weary and my very soul is being
crucified. I am jus bearing tho but I can not go on bearing much longer.
You see it is terrible, maybe a curse is upon me or something is the reason
why I should go on suffering in such way. I say again that it is horrible to
think that jail houses is killing me. . . . This longest stay incarcerated is

becoming to be a low death for me. And the rest of the boys are likewise slowly dying too. Anna I am lonely. There is no one to be my comforter.[30]

Patterson is quite eloquent and thoughtful about the issue of his sanity, returning to the subject of his picture in the *Labor Defender* as one source of his depression:

I think I am entirely sane of course you people in the East may not consider it like an I imagine. I suppose when you have completed this letter and carefully consider all things you are very likely to see things differently too. . . . Brooding over this misfortune and troublesome situation have actually got a lots to do with one mental an especially when I hear of how I am being slandered. Its made me actually foolish. . . . of course I am sane in my own way, but being tortured continual by such an being Scandalized. Keeps me in rather some what bitter moods. You know that when especially my person is impossible to do anything at all about it. And those photos being publish of me have really became outragous and indecent remblance [resemblance] which they have resulted. And I beg of you all to cease publishing them. I disapprove such Horrible looking things. And it is Hurtful and Dispirited to me.[31]

Anna Damon's immediate, heartfelt response signaled a new stage in their sometimes testy and contentious relationship:

Dear Haywood:

I just got your last letter and I cannot begin to tell you how deeply it moved me. In one way it made me very happy to realize that you feel that I am such a real friend to you that you can pour out your heart to me the way you did. You are a very brave and fine young man, Haywood. You have stood up under your terrific burden in a way that very few older people would have been able to do.

But really, Haywood, you must not brood so much about your mind. Anyone who just reads your letters can see that you have a splendid mind. . . .

As for slandering you—you must give up that idea, Haywood. I agree with you about the picture, but honestly, it is the only one we have of you and we would be only too glad to use another one if we had it. Perhaps you can have a picture taken of yourself. We will try to have it printed all over so that the people will see you as you look today. There are thousands of people who would like to see it.[32]

Patterson responded with a similar depth of feeling:

My Dear Miss Damon I was surprisingly pleased at having received such prompt favorable response from my last letter to you. I am supremely pleased to say that I have read your nice letter through several times, with a great deal of pleasure. . . . and right now I am utterly unable to tell you how sincerely I feel for you Anna. But yet you seem to doubt my confidence I have in you and I suppose you have good reasons for doubting. And that comes largely on account of my conduct some times. I say things I ought not to say. But allow me to reassure you that I have certainly taken you into my confidence in so personal matter. And have liked and trusted you. And you know that. Regardless to my pretends to be angry some times at you all why I hardly be anyway angry at all you see? Of course Anna I feel always that you are a real friend whom I can confide in. I read in your eyes last time I saw you. I saw there in them good things.[33]

Never one to miss an opportunity, however, Patterson seizes this moment to lobby for a camera:

Please Anna I have not got anyways of having any pictures made of myself right now, but if you promise to cease using the ones you all have got I will try my uttermost to get some done when the weather get more pleasant here. You see I have been planning to get me self a Kodak, but I haven't had the money to spare for such, you see. I can get a pretty good Kodak for $5.75 and no soon then I have the money to spare I shell get myself one. And then you can rest assure of getting a picture every time I have some made I promise.[34]

Still, there is no question about his sincere regard for Anna Damon:

Well I can have nothing more to say at this time except to say good bye until I hear again from you. And please don't ever think I am tired of you. I don't get enough of you.[35]

It is impossible to know what happened behind the scenes in the ILD office, but shortly after this exchange, Patterson received a jokingly flirtatious letter from Rose Baron of the ILD:

Dear Haywood:
I feel a little jealous of all the letters you have been writing to Anna Damon without sending any to me. You know how anxious I am to hear from you all the time.[36]

Patterson was at his most courtly and flirtatious in his immediate response:

My Dear Miss Baron I thank you for your generous kindness and courtesy an I have just received your most welcome letter containing the regularly monthly allowance. Now may I reassure that I don't want you to ever be conscious of a thought or never imagine that I am forgetting you even for a moment because I am not. And in fact why I while a way most of time sitting around thinking mostly of Anna and you. . . .[37]

From this point forward, Anna Damon apparently shared the task of corresponding with Haywood Patterson with Rose Baron and Hester Huntington. Patterson held Damon in high esteem and regarded her with genuine affection; when he learned about her death, he was truly grief-stricken:

My Dear Mrs Huntington.

I have received your most disheartened letter with telling me of death of our dear beloved friend Miss Anna Damon. Now I can assure you that to me is one of the toughest blows any man can possibly stand. You see Miss Damon was to visit me twice and I promised her that I would some day meet her a free man. But what a disappointment god Have made between us. Now I feel Heartbroken and my grief is to great to express. I feel as if I had losted a Brother or Sister Because Miss Damon was very dear to me all the while she gave of her best assistance toward helping we boys. I can never forget Miss Damon and she must live again in our Hearts & Minds. We must speak of her often. Now my friend Ozie says that he regret the death of Miss Damon. Now my heart is too full to put into words on paper my regret over Miss Damon death.[38]

After the Alabama authorities dropped the charges against Eugene Williams, Olen Montgomery, Willie Roberson, and Roy Wright in July 1937, but sustained Patterson's fourth conviction and sentence of seventy-five years, Patterson was sent to Atmore State Prison, where he would spend the next six years of his life. Charlie Weems was paroled in 1943, Andy Wright and Clarence Norris the following year.

Writing to Kay Boyle's daughter Bobby from Jefferson County Jail in Birmingham, where he was recuperating from a leg injury, Haywood rambled on about Bobby's pet frogs and her pet dog before turning inwards:

Since my last Writing you I Have been confined in an infirmary suffering with one of my leg. I have written Mother Vail [Kay Boyle, who was then married to Lawrence Vail] about it. . . . I spent a month and two weeks at the infirmary at Kilby Prison, Montgomery Alabama. . . . and then after I gottin almost well I was transferred to Atmore Alabama which is a state prison farm. I spent two weeks there working Before I was brought Here. The desolate stretch of land & woods where Atmore Alabama is situated is a very lonely place. I didn't like there at all. . . . I also explained in mother Vail letter about four of the Scottsboro boys Being freed. And we other five are looking forward for our liberty.[39]

In the meantime, as hopes for his release faded, Patterson continued to pursue more elaborate—and desperate—money-making schemes through his far-flung correspondence.

One such scheme involved someone named Maria Sulek, who Patterson apparently had approved to raise funds on his behalf:

My Dear Miss Maria Sulek. And to all whom this may concern. I am writing Miss Sulek because she have seemed so willing to cooperate with me to the fullest extent. And what more she understand the situation and may I now express my belief and confidence that I have in her. Now I am most certain that you all know about all the boys getting released except myself and Powell. Of course it was hardly fair to see all the boys go and I remain on incarceration. Well Dear friends I was just turned out to be the goat, but to me I have done everything I could to obtain my freedom. And success seems as far from me now as before. Some how the people says my prison record wasn't good enough. But I have tryed very hard to be good but there were those who had personal grudge against me. Why I have bent over backward to decent [*sic*] to everyone. Now my main reason for writing you dear workers at this time you see it have come a time that I should ask for some very needed assistance. You see dear workers I have consult a lawyer of this State and He have assured me that He could get me released soon for the amount of $500. Now my sister who is poor recently sent me 200 and $50 and I now need 200 and 50 more dollars before the lawyer can act. Now he do not wish the money until He have done the work. Now I do Hope and trust you all will raise whatever you can of this amount. Your friends can give from one cent up because every cent would be necessary. So please send me whatever you all can possibly raise within the next 3 weeks to come. Now it is you all duty my friends to exert every effort to get me released at once because I can not endure the suspense much longer. My long confinement in prison Have done a lot to my neverous system and Health condition and if I don't soon get out of this climate I am to die a slow Nouseous death.[40]

Apparently, Patterson would routinely send out letters soliciting funds to randomly selected individuals and organizations and they, in turn, would seek advice from the ILD about how to proceed. Dick Cardamone of the CIO-affiliated American Communications Association wrote the ILD:

Several days ago I received another letter from Kilby Prison ... from a chap named Haywood Paterson asking for financial assistance. Apparently from his letter Kilby Prison is not too good which of course we can readily believe.

The purpose in writing you is to get some information on this chap. Is Haywood Paterson one of the Scotsboro boys? If we hear favorably from you we will make a small contribution but I am wondering whether we should send it to you or to the boy direct.[41]

Don Jonson wrote from La Crosse, Wisconsin to express his concern about a request Patterson has recently made of him:

I am troubled about our friend Haywood Patterson and hope you won't mind my turning to you for help. He implores me for a sum of money with which to secure his freedom. He doesn't make clear to me how this is to be accomplished and I feel too uncertain of the matter to go ahead. In his latest letters he says you have sent him $100. to be used in this purpose and he begs me for $75. more. I am far from being wealthy, but would scrape up this amount for him if convinced it would bring him his release. Unfortunately, I've already tried to help other prisoners with cash and in both cases my aid turned out illadvised. These men took advantage of my interest and deceived me.

I don't understand, if Haywood's letters are censored at the prison, how he can write as openly as he does; and at that I cannot always be sure of his meanings. I keep sending him money—usually $5—every month, but now he wants far more.[42]

Responding on behalf of the International Labor Defense, Louis Colman, in effect, called Patterson a liar and—perhaps reflecting a growing impatience with his tactics—warned Jonson away:

I am afraid that Heywood is not informing you exactly of the situation. We have not sent him any sum of $100, and have no intention of doing so. We send him $5 a month, which is about all he is able to spend for legitimate purposes in prison.

Unfortunately he has been able to secure considerable other sums through a well-organized campaign of letter writing, and as a result has gotten himself into trouble over and over again in the prison. There is no doubt that the scrapes he gets into in the prison through such spending have contributed to the difficulty of obtaining his release.

You are correct in saying that it is ill-advised to send him any large sums of money.

Please consider the contents of this letter confidential, as Heywood's prison psychoses make it difficult to understand why people should not send him money.[43]

Colman's view of Patterson's behavior exactly echoes that of Allan Chalmers, head of the Scottsboro Defense Committee, who in an earlier letter to Hester Huntington ruminated about Patterson's psychological condition:

In the letter to me he dwells at great length upon the quite natural feeling of "treacherous enemies." It is what the psychiatrists would consider an evidence of a persecution complex. I do not feel it, myself, to be necessarily so dangerous that he could not be rehabilitated very successfully on release. I am much more worried about the other psychological and physical habits which are so natural to prisoners that they call them "prison psychoses."

If it were not for the fact that a proper psychiatric examination would be helpful in our attempt to get his release, I would not, of course, adopt that method, although I do think, in any case, we have to be realistic enough to face the situation as it is, even though we agree, as you and I do, that he is not responsible for the condition but is a victim of it.[44]

This is a view that Anna Damon would not have supported, publicly or privately, and Chalmers's use of the language of psychotherapy in relationship to Haywood Patterson signals a subtle but important shift in the ideological perspective of his official defenders. As the Scottsboro case began to disappear from public view in the mid-1940s and as the International Labor Defense receded from the political landscape (becoming reconstituted, along with two other organizations, as the Civil Rights Congress in 1946), the emphasis on the remaining Scottsboro defendants seemed to shift from their social and political victimization to their personal and psychological characteristics. From this perspective Patterson's *Scottsboro Boy* can be seen as a strategic attempt to reawaken the public passions about the Scottsboro case that had greeted it in the 1930s.

⁙⁙⁙⁙⁙⁙⁙⁙⁙

The product of a sustained dialogue between a shrewd, wily, pragmatic survivor—albeit one with a deeply embedded sense of fairness and justice—and a veteran journalist and committed Communist, *Scottsboro Boy* presents a somewhat different picture of Haywood Patterson than the one revealed, in various guises, through his voluminous correspondence. In his collaboration with Earl Conrad, Patterson invariably depicts himself as a figure of staunch, unwavering resistance to injustice and racial oppression, sometimes in contrast to his co-defendants—who he describes as less capable of bearing up under the strain:

> We heard them yelling like crazy how they were coming in after us and what ought to be done with us. "Give 'em to us," they kept screaming, till some of the guys, they cried like they were seven or eight years old. Olen Montgomery, he was seventeen and came from Monroe, Georgia, he could make the ugliest face when he cried. I stepped back and laughed at him.[45]

On several occasions, when the boys take collective action in the early days of their imprisonment, Patterson inevitably casts himself as their natural leader:

> We didn't like nothing at all about the place; we didn't like our death sentence; and we decided to put on a kick. I said to the man who brought me a prison meal, "I don't want that stuff. Bring me some pork chops."

"Huh, pork chops?"

"Yes, pork chops. You got to get it. We're going to die and we can have anything we want."

All the fellows put down a yell, "Pork chops!"

We crowded up to the bars. We put our hands out and shook fingers at him. We hollered, "Pork chops. Nothing else."[46]

In another episode the black inmates in Birmingham jail strike to protest their treatment on visiting day:

We put on a terrible strike. We didn't allow nobody in our day cell. We hollered and stomped and we wanted our rights, we put up such a howl the sheriff of Jefferson county, he came in with some trusties and guards. They wanted to beat on us. The sheriff wouldn't let them. He pulled them away. From prisoners in our sympathy we had got knives, iron rods, everything to fight with. I was ready to fight.[47]

A standoff follows and the prison officials try to starve the convicts into submission. Patterson makes the strategic decision to bring the strike to an end:

We went for almost two days without food.

After we got weak I decided we should give up. The prison people, they weren't going to give us food. If we'd have starved to death we'd have done them a favor. We were hungry. I told the boys we better try to make a deal. They would mostly follow my mind.[48]

He negotiates a deal with the warden—a very modest one by contemporary standards—whereby the convicts agree to relinquish their weapons in exchange for fair treatment in the jail:

"We want visitors. We also want to be let out for exercise twice a week like the others. We be willing to pay for the damage we did to the cell block. . . ."

Good food we got right away. The warden sent up candy which we could buy.

Things changed. We saw visitors. We got exercise like the others.

Without that strike we wouldn't have had anything.[49]

Patterson doesn't hesitate to underscore his importance in these struggles: "The jailors, they believed I was putting ideas in the other eight boys' heads. They figured I was teaching them how to rebel."[50]

His accounts of collective struggle by the prisoners include one striking episode of interracial solidarity at Atmore State Prison Farm—an unmistakable salute to the contemporary interracial militancy of the CIO:

We didn't have too much contact with the white prisoners. Their squads were all white, ours all black. They'd be working somewhere, we'd be working somewhere else. Sometimes though we would be close by them.

About four white squads and two Negro squads, we struck once over the bad food they were sending out. . . . A signal passed between the white squads and our work gangs. About a hundred of us altogether, we just squatted. We told our guard, Captain Landers, the cussing captain, we wasn't eating, and we wasn't working. . . .

Captain Landers, he got going on us. "You giving us a sit-your-ass down strike? Maybe you want collective bargaining? Maybe you want the C.I.O. set up the Atmore local of the Farmer's Union? *Maybe you want to get your heads shot right off your goddamn necks?*"[51]

Episodes such as these—highlighting as they do the power of collective consciousness and collective action as the most effective means of combating racial and social oppression—constitute the heart of good political fiction and memoirs, but these images of the Scottsboro Boys as a collective force occur infrequently in Patterson's autobiography. As his personal correspondence makes clear, the boys were distinct individuals with their own personalities, values, and beliefs. Factions formed and re-formed; they argued and sometimes fought among themselves—and Patterson was often cast as the outsider of the group. Accounts of collective struggle and images of the Scottsboro Boys as symbols of resistance were difficult to sustain under these conditions. What emerged, in fact, in *Scottsboro Boy* was a portrait of Patterson's stubborn, proud, defiant, and lonely individualism—values which, interestingly enough, Richard Wright was criticized by some Communist Party officials, like Benjamin Davis, for emphasizing in his fiction over the ideals of mass struggle.

If *Scottsboro Boy* departed from the values and conventions of politically committed writing in its emphasis upon the triumph of the individual will over adverse circumstances, it transgressed even further in its graphic treatment of the issues of sexuality and criminality—a portrayal unequaled in twentieth-century American prison literature up until that moment.

The question of how to satisfy one's sexual needs was a persistent undercurrent in the correspondence of many of the Scottsboro Boys. Patterson spoke openly in *Scottsboro Boy* about how the money they received from their supporters made it possible for them to purchase virtually anything inside Jefferson County Jail in Birmingham, but he was very discreet about sexual matters in his correspondence with Anna Damon. Olen Montgomery, on the other hand, was blunt and direct:

Listen Anna

This is a matter of business no fun. Now listen you realize how long I have been cut off from my pleasure don't you? And you realize it had on me don't you? Especially me being a young man. An you or anyone else should have sympathy and feeling for me. Now listen please don't get offended over what I am fixing to say in your present. I am almost crazy in this place. This is what I want to say it's a new warden here at night and he will let me to a woman for all night for $5.00 which is worth it. I know its worth $5.00 of mine. Because if I don't get to a woman it will soon run poor me crazy. I really have stood it long as I can. I just got to get to one. I have been in jail over five years. And it's a shame. Now Anna please send me $5.00 and take your out of my eight. Now please send it rite away I am looking for it Friday. And I will appreciate that the very highest. Now since I have explain the situation to you. I am sure you will feel my sympathy. Want you? Sure you will! I am looking for it Friday without a fail. Please give my regard to all.[52]

Opportunities for sexual encounters with women convicts, such as those presented at Jefferson County Jail, were nonexistent at the notorious Atmore State Prison. There, Patterson recalled, male convicts were faced with starker choices:

Before I got here I heard tell of the gal-boy life at Atmore. Gal-boy stuff went on at Kilby and at Birmingham jail too. All prisons all over, I guess, but I wasn't interested in it. I was a man who wanted women. But this was the main thing going on among the Atmore prisoners. . . .

Old guys, they called them wolves, they saw me looking at this stuff and thought I might be a gal-boy. One came up and propositioned me. I didn't like that none at all. I said, "If any of that stuff goes on with me in on it I'll do all the fukking myself. I been a man all my life." I doubled up my fists to show him I meant it.

Guys kept coming up to me. Saying no didn't stop them. I was being forced to be either a man or a gal-boy. They wouldn't hardly let you stay outside of this life. I would say, "Leave off. I'm looking for a gal-boy my-self." That was a hard-guy act I was putting on at first because at first I didn't look for no gal-boy.[53]

Under constant pressure, however, Patterson relented, becoming, in the parlance of Atmore, a "wolf" who fought as ferociously as the other "wolves" to protect his sexual interests.

Patterson was equally graphic in his description of the various rackets that constituted the economy of prison life. Always the survivor, he carefully calculated the best ways to raise the funds he needed:

The outside world, it became less interested in the Scottsboro case. I still heard from people, but less money came to me. I had to look for ways to make money in prison. There were plenty. The place was loaded with petty rackets. I was in on many of them. . . . I couldn't eat the regular mess. I had to have money to buy decent food. I had to take care of my gal-boy too.[54]

Patterson's clinical anatomy of the intricate, multilayered connections and transactions among convicts and the prison administrators—and his harrowing accounts of the depths of human depravity he witnessed and encountered, among prisoners and officials alike—effectively demolished bourgeois expectations about the prison system as an effective means of social and personal rehabilitation.

Patterson recounted his story with a graphic sense of recall, a relentless accumulation of brutal detail, and an understated, acerbic sense of humor. He offered this view of the courtroom upon hearing of his first conviction: "That courtroom was one big smiling white face." He could sum up the psychosexual dynamics of southern race relations in a sentence: "Down there, the way they figured, a black man and a white woman fifteen hundred miles away was still too close."[55] Or wryly express his sense of the capacity of the southern black community to organize itself against racial oppression: "Negroes down South, it is hard to get them to see things right, whether the free ones or those in prison. They don't have enough stand-together feeling where whites are concerned."[56] Or offer succinct commentary on the nature of southern justice: "Color is more important than evidence down there. Color is evidence. Black color convicts you." And, finally—flowing from this assumption—offer his pragmatic recognition of the value of interracial struggles for social justice: "If a white person says you did something, you did it. There's only one way a Negro can get out of it when this kind of evidence, *white testimony*, is brought in. That is to have rich white folks or labor organizations fronting for you . . . What happened in the Scottsboro case wasn't unusual. What was unusual was that the world heard about it."[57]

As one man's account of a flagrant social and political injustice and his descent into the absolute depths of a human bestiary, *Scottsboro Boy* was riveting and unforgettable. Still, its claims upon the social and political imagination of its readers were somewhat ambiguous. At the end of his journey Patterson tried to sum up its meaning, in language that seems to replicate the rhetoric of the Civil Rights Congress and its predecessor, the International Labor Defense:

Those people . . . who fought for me, I am grateful to them. To all who helped in the courts, wrote letters, sent money, picketed and marched and died for us boys. They helped me, helped the country, helped my people. I guess my people gained more off the Scottsboro case than any of us boys did. It led to putting Negroes on juries in the South. It made the whole country, in fact the whole world talk about how the Negro people have to live in the South. Maybe that was the biggest thing of all. Our case opened up a lot of politics in the country. People said more about lynching, the poll tax, and a black man's rights from then on.[58]

But this broad, sweeping rhetoric also contained within it a more poignant and personal vision of America's possibilities:

Sometimes as I lay out here in the North in my little room waiting either for the law or freedom to come and take me, I think of my people down South. Then I want to go there, be among them, live there. I think of a small town maybe where I might settle, and have a home and a family, maybe a business or a small piece of land. It's what a man needs. What my people need. Land. The land they live on, have worked so much and owned so little.

But I won't have that unless the people say so. The people must bring an end to the Scottsboro case once and for all.

They must say whether I suffered enough—or whether I go back there to be tortured to death. That's what they'll do with me, for sure, if they get the chance.

I have had a great struggle. But I want the world to know I am unbeaten.

I'll lay out here in my room—and see what you do.

Then I'll make my next move. . . .[59]

Containing faint echoes of the old Communist Party slogan "Self-Determination in the Black Belt," the final words attributed to Patterson were nevertheless a striking admixture of tense expectancy and fatalistic acquiescence, idealism and pragmatism, defeatism and hopefulness, passivity and defiance—reflecting in the final analysis a vision of social incorporation rather than one of radical political transformation.

The publication of *Scottsboro Boy* in June 1950 coincided almost exactly with the release of the last Scottsboro Boy still in prison, Andy Wright, who won his second parole after being re-imprisoned on an earlier parole violation charge—signaling the symbolic closure to a spectacle that had now dragged on for almost two decades. Events did not transpire so neatly in the real world. Had Haywood Patterson been more attentive to the tactics many nineteenth-century fugitive slaves used to cover their tracks, even as they offered their lives to public view, perhaps his life might have taken a different trajectory. Shortly after its publication, on June 27, 1950, as Patterson was leaving the office of the Civil Rights Congress in Detroit, he was arrested by the FBI and held on $10,000 bail for fleeing Alabama to avoid imprisonment. Alabama authorities demanded his extradition. The Civil Rights Congress fought to have bail reduced, put up the money for it, then organized a nationwide letter-writing campaign on Patterson's behalf. Alabama authorities demanded

Figure 6.4 Haywood Patterson, left, in Detroit on July 13, 1950. At right is Patterson's attorney Ernest Goodman, executive of the Civil Rights Congress. Source: AP Images.

his extradition, but Michigan Governor G. Mennen Williams stood firm in his refusal to return him to Alabama. This convergence of events doubtlessly heightened public interest in *Scottsboro Boy*. For many readers and critics, however, it was yet another opportunity to replay the old debates about Scottsboro, with many of the same cast of characters and many of the same arguments, complete with the usual insinuations about the motives of the Communist Party.

Independent leftist journalist I. F. Stone hailed *Scottsboro Boy* as "one of the great documents of the history of the Negro race in America, and of man's endless inhumanity to man," and celebrated its potential for bringing about social reform in the United States:

> Unless the conscience of America is dead, it gives all of us a fighting chance to clean up the prisons and the slave labor camps in which white man and black are held in awful bondage in Alabama and other Southern, perhaps some Northern states. . . .
>
> We ask ourselves—how could the Germans allow the Nazis to do what they did. Let us ask ourselves—how can we allow Southern white men to do to Negroes and other white men what this book reveals? This is our shame. This is our test.
>
> Two tasks are made urgent by this book. One is to win a pardon for the last of the Scottsboro Boys. . . .
>
> The second task set for all of us who claim to care about human values is the cleanup of the prisons and prison camps in which human beings, black and white, are driven liked doomed creatures on slops of food the pigs turn away from.[60]

He was joined in his enthusiasm by the black Harlem-based Communist Abner Berry, who called his own review of Patterson's narrative "American Dachau."[61]

Still, the reverberations of the historic conflict between the ILD and the NAACP continued to inform reviewers' responses to *Scottsboro Boy*, and Patterson's indisputably pro-ILD sentiments rekindled the resentment of the NAACP and some mainstream liberals.

Writing for the *New York Times*, J. N. Popham characterized the book as

> a piece of primitive writing and thinking based on an accumulation of so much evil among men that one finishes it with a kind of reader's trauma. . . . It is not a nice story. It is a monolith of human degradation. His report on the sex life of prison inmates and the beatings administered by prison guards is sickening. His thought-processes are extremely one-sided, and as the years of injustice roll by he adopts completely the hardened criminal's attitude that only he can really understand the maddening imbalance of

the scales. The reader soon begs for some touch of logic to Patterson's depiction of human beings, no matter what their depravity.[62]

And Hodding Carter's review for the *Saturday Review of Literature* resurrected some of the old accusations and debates surrounding the Scottsboro case:

> It is understandable that a Negro victim of white malice should not try to analyze the whys of his long mistreatment. His story is, on the surface, only an ugly account of unrelieved evils. . . . But it is not understandable from the text why Haywood Patterson, whose memory for long-ago incidents even to degraded and inconsequential prison small-talk seems uncanny, should overlook an initial reason for the State of Alabama's adamant demand for an undeserved penalty. . . . Throughout the sordid narrative Patterson credits the International Labor Defense with giving him the only effective help. Never does he even name the National Association for the Advancement of Colored People. . . . Yet Walter White himself went down to hate-charged Scottsboro as soon as the story became known. The NAACP employed the defendant's first counsel.[63]

Carter's judgments were endorsed by the indefatigable Walter White, who dismissed *Scottsboro Boy* as ILD propaganda.[64] Echoing these views, the staunchly anti-communist head of the Scottsboro Defense Committee, Allan K. Chalmers, resurrected the long-standing charges about Communist cynicism and manipulation:

> The fact remains that after he discovered himself to the [Civil Rights Congress] and *Scottsboro Boy*, the story of his imprisonment was published, the FBI came and arrested him. They had to. His book put them on the spot.
>
> They did so as Patterson walked away from the Detroit office of the Civil Rights Congress. Under the circumstances, the arrest was a result either of criminal neglect, incredible stupidity, or deliberate intention on the part of the CRC, an organization under the constant surveillance of the FBI.
>
> If Patterson did not know that, he should have been told. But the logic of the Communist point of view has it that Patterson is a more valuable asset in jail than out. . . . Behind bars, he could be used as an excuse for gathering funds to pay his bail, as an example of America's inadequate notion of social justice and what not. Whether they meant to or not, they did use him again.[65]

As another dedicated anti-communist, William A. Nolan, succinctly put it, the Communists "hoped to drain the very last drop of Patterson's blood."[66]

Thus, in the early years of the Cold War, the Scottsboro case still had the capacity to arouse fierce partisan debates. For Haywood Patterson, however, these arguments would become increasingly abstract and, ultimately, moot.

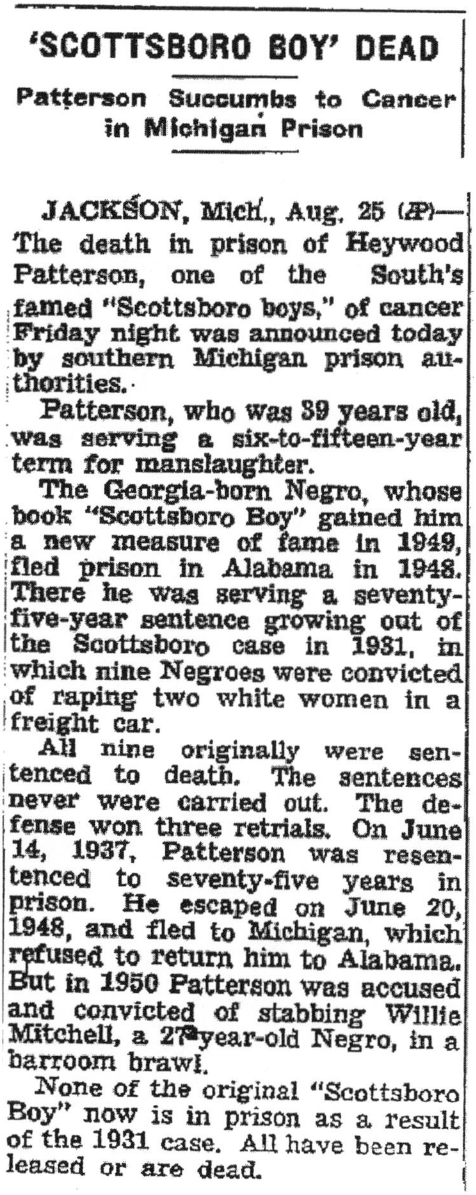

'SCOTTSBORO BOY' DEAD

Patterson Succumbs to Cancer in Michigan Prison

JACKSON, Mich., Aug. 25 (AP)— The death in prison of Heywood Patterson, one of the South's famed "Scottsboro boys," of cancer Friday night was announced today by southern Michigan prison authorities.

Patterson, who was 39 years old, was serving a six-to-fifteen-year term for manslaughter.

The Georgia-born Negro, whose book "Scottsboro Boy" gained him a new measure of fame in 1949, fled prison in Alabama in 1948. There he was serving a seventy-five-year sentence growing out of the Scottsboro case in 1931, in which nine Negroes were convicted of raping two white women in a freight car.

All nine originally were sentenced to death. The sentences never were carried out. The defense won three retrials. On June 14, 1937, Patterson was resentenced to seventy-five years in prison. He escaped on June 20, 1948, and fled to Michigan, which refused to return him to Alabama. But in 1950 Patterson was accused and convicted of stabbing Willie Mitchell, a 27-year-old Negro, in a barroom brawl.

None of the original "Scottsboro Boy" now is in prison as a result of the 1931 case. All have been released or are dead.

Figure 6.5 "'Scottsboro Boy' Dead." Source: New York Times, August 26, 1952.

In December 1950 he was arrested in the aftermath of a barroom brawl that led to the stabbing death of another man. Charged with murder, he claimed self-defense—claiming in one version of his defense that he had been in the bar selling copies of *Scottsboro Boy*, but had not been involved in the brawl—and stood trial three times. The first one ended in a hung jury, the second in a mistrial, and the third in a conviction for manslaughter. Sentenced to six to fifteen years, Patterson was diagnosed with cancer shortly after his last trial. Less than a year after he was imprisoned in a state penitentiary in Michigan, he died on August 24, 1952 at the age of thirty-nine.[67]

||||||||||||||||||||||||||||||||||||||||||||||||||||||||||||||||||||||||||||||||||||||||||||||||||||||||||||||||||||||||||||||||||||||

CHAPTER SEVEN

COLD WAR SCOTTSBOROS

IT WAS NO ACCIDENT that the Scottsboro Narrative that flourished from the early 1930s until the late 1940s roughly coincided with the rich period of American cultural and political life Michael Denning examines in rich and meticulous detail in *The Cultural Front*: "the extraordinary flowering of arts, entertainment, and thought based on the broad social movement that came to be known as the Popular Front"[1] — or that its demise, or, more precisely, its transformation, coincided with the decline and fall of this movement, casualties of the Cold War, post–World War II Red scares, and widespread anticommunist purges.

Don Mankiewicz's 1955 Harper Prize–winning novel *Trial* signaled a version of the Scottsboro case more in keeping with the social and political climate of America in the 1950s. Set largely in a small town in a west coast state that seems suspiciously like California, even as Mankiewicz goes to great lengths to indicate that his fictional setting is *not* California; apparently inspired by the Scottsboro case, even as Mankiewicz translates the salient details of the case into a significantly different setting and racial-cultural dynamics; *Trial* offers a particularly graphic example of the distortions of the imagination and the moral and political evasions characteristic of American life during the peak years of McCarthyism.

On a June night in 1947, Angel Chavez, a seventeen-year-old Mexican American, a high school senior, wanders onto San Juno Village Beach, drawn by the ritual of the first grunion run of the season. Angel, who lives in "Mex Town," right outside of the village boundaries, thinks that he is trespassing — even though the novel makes it very plain that he is mistaken in his beliefs:

> Actually — although Angel did not know this — there was no law that barred him from becoming a resident of the village, and the State — which, having had two successive progressive administrations, had both a "little FEPC" and an antidiscrimination act — stood ready to enforce his right to live wherever he chose. He had only to enter the office of any real-estate broker who handled village property, obtain proof that the broker had property inside the village line listed for sale, deposit the price of the property, and then prove that the broker's refusal to sell to him was motivated solely by Angel's Mexican descent.[2]

In this way readers are positioned to witness the circumstantial and accidental nature of the events that trigger off the subsequent events of the novel. But the narrator's essentially legalistic description of Angel's circumstances—apparently delivered without a trace of irony—also reveals Mankiewicz's bedrock faith in American values and institutions, particularly its sense of fair play. These values underscore the story he tells in *Trial*. Thus, the narrator editorializes in a dismissive fashion about Angel's fear of being discovered on the beach:

> The fact of the matter is that Angel's terror was largely unjustified. Mexicans were frequently found on the beach. Usually they were migrant workers from California to the south; often they could not read at all, and were genuinely bewildered when challenged. . . . The worst that might befall a Mexican caught on Village Beach, in cold fact, was that while he was being ejected, chivvied back to the sea wall, and up the steps to the road, he might overhear an occasional insult.[3]

As Angel climbs the darkened steps from the beach, he has a chance encounter on one of the landings with Marie Wiltse, the fifteen-year-old sister of his classmate and friend, Stretch. Subject to severe attacks of rheumatic fever, Marie is beset with demons of her own. Like Angel, she feels that she does not belong on the beach either. Her sense of their shared alienation leads quickly to an awkward, adolescent petting episode which ends tragically when Angel accidentally tears the top of her dress. Marie breaks away, runs for the steps, collapses, and dies. Angel calls for help, a crowd is quickly drawn to the scene, and Angel is arrested by a Fish and Game Commission authority "to prevent a lynching," as the commissioner later explains—even though the narrator interjects that such an incident, "had it occurred, would have been the first in the State in three decades, and the first since the era of 'Chinese cheap labor' if lynching is considered to include a racial factor."[4]

The protagonist of *Trial* is David Blake, a thirty-year-old Assistant Professor of law whose teaching contract is jeopardized because he lacks practical legal experience. His mentor, Terry Bliss, intervenes with the Dean of the law school and persuades him to give Blake a summer's grace to acquire some of the necessary experience. Fortuitously, Blake comes across the law office of Barney Castle, who just happens to be on the verge of assuming the responsibility of defending Angel Chavez. Blake eagerly agrees to work with him. Barney Castle, his assistant Abbe Klein, and Blake descend upon the San Juno jail to interview Angel. Along the way Castle explains to Blake the difference between his approach to the case and that of the Mexican Advancement Association (MAA), in terms that would be immediately recognized by anyone familiar with the history of the Scottsboro case, particularly the long and contentious relationship between the ILD and the Communist Party and the NAACP:

"A bunch of handkerchief heads," Barney growled, answering David's inquiry into his reason for not wanting them to have the case before David could ask it. "They'll tell you themselves that all they want is to see the boy get a fair trial. That's their policy. If he's guilty, they'll want him found guilty. If he's innocent, they want an acquittal. That's not enough. Not for a kid in the spot he's in."

"Why not?" David asked innocently.

"Why not? . . . Fairmindedness is the judge's job. . . . Maybe he sticks to it, maybe he doesn't. The prosecutor's job is to get his conviction. And, chances are, he has plenty of help. . . . Look here. . . . Here's the poor Mexican. Here's the prosecutor, straining his gut for a conviction. Up above it all is a fairminded judge, keeping order. Doesn't it need something else to give that boy a chance? You bet it does. . . . It needs somebody working the other side of the fence, somebody who doesn't care about fairness any more than the prosecutor, somebody who wants to get the boy off just as badly as the prosecutor wants his conviction. And that's just what the M.A.A. won't provide."[5]

Within minutes of his meeting with the MAA lawyer, Jiminez, in the San Juno jail, Castle peremptorily dismisses him, and Jiminez—unlike the leadership of the NAACP in the Scottsboro case—meekly accedes. Mankiewicz also makes a calculated decision to make the presiding judge of the case an African American—almost as if he wishes to dramatize the progress that has occurred in the United States since the infamous Scottsboro trials of the 1930s. Still, the presence of Judge Theodore Motley allows Angel to question how secure his status really is:

"One of the men at dinner last night," Angel said. "He told me that the reason it will be a colored man is because they can trust him to do his best to hang me—on account of his not being white either. Is that true?"

David hesitated before replying. In his judgment, Angel had stated rather precisely why Theodore Motley had been assigned to hear this case. It was David's view that the intent of those who gave him the assignment was that he should feel that he himself was, in part, on trial, that not merely the guilt or innocence of the defendant was to be tried, but also the question of whether a dark-skinned judge could dispense even justice to a slightly less dark-skinned defendant.[6]

At another moment David notes Judge Motley's manner of speech:

Even in those few words, the Judge's clipped, vaguely Continental accent was apparent. The accent was because he had been educated abroad. This, David thought, was because it had been difficult for a would-be judge whose skin happened to be black to get his undergraduate education close to home. Yet now, David's thoughts raced on, people rose and were silent when this man came into the room; and his career was sure to take him far, unless, by some chance, Angel Chavez were to be acquitted.[7]

Moments such as these in *Trial* have the effect of underscoring how the historical circumstances of the Scottsboro case are both displaced and reinvented by Mankiewicz's imagination. But to what end?

Once David Blake enters the Chavez case—or, more precisely, once Blake is permitted by Castle to defend Angel—*Trial* becomes a legal procedural thriller of sorts, one that pits Blake's idealism and innocence against both the power of the state and Barney Castle's craftiness and political cynicism. In this context the relegation to the background—or the complete disappearance—of some of the key issues historically associated with the Scottsboro case allows Mankiewicz to highlight the issues central for his purposes. In Mankiewicz's hands, Scottsboro functions as a springboard for an investigation of the crises facing American society in the mid-1950s. Indeed, the publication of *Trial* coincides almost exactly with the publication of the results of a public opinion poll, administered to "more than 6,000 men and women, in all parts of the country and in all walks of life" by the American Institute of Public Opinion and the National Opinion Research Center from May through July 1954. The survey was designed to probe in some detail American reactions to "two dangers. One, from the Communist conspiracy outside and inside the country. Two, from those who, in thwarting the conspiracy would sacrifice some of the very liberties which the enemy would destroy."[8] Like the pollsters, Mankiewicz is concerned with probing, but in fictional terms, the implications of these two dangers for contemporary American life.

For his critique of the first danger, the Communist conspiracy, Mankiewicz freely draws upon a repository of stock anti-communist stereotypes widely available in 1950s American popular culture. Abbe Klein, for example, is a version of the Communist *femme fatale*, who uses her seductive wiles to dupe innocent American men like David Blake. Although Barney Castle pointedly defines his relationship to Abbe—"I call her Miss Klein by day, and Abbe by night"—Abbe initially functions as an effective assistant and wise counselor to Blake. Within weeks, however, Abbe and David have become lovers, and Abbe, on several occasions, reveals herself as someone who is capable of making questionable, even unethical, legal judgments. She argues against David's petitioning for a change of venue for the trial, even though the public passions aroused by the Chavez trial would seem to support such a request; and she strongly hints that the information David uses to force the dismissal of a jury panel would have had more impact if he had suppressed it until after the trial—at which point he could have made a strong case for a mistrial. The way both of these issues are handled has some impact on the outcome of *Trial*, although not necessarily a decisive one, but it is only two-thirds of the way through the novel that readers learn the real basis of Abbe's shady ethics:

she is a Communist, or she was one until the moment that she reveals her past to David. In glum words that echo the narratives of seduction and betrayal recorded by former Communists in post–World War II works like *The God That Failed*,[9] Abbe recounts the story of her journey to the light:

> I was [a Communist] for a while. A short while. When we were leading the workers of the world in the fight against fascism. Then came the Hitler-Stalin pact, and the fight against fascism became an imperialist war. It was tough for me, but I made it. Then Hitler broke the pact, and the imperialist war became a holy crusade, and that was one switch too many."
>
> "You got out."
>
> "Out of the Party," Abbe said slowly.[10]

Since that time, Abbe reports, she has been moving, gradually but inevitably, towards a final break with the Party—a process, she wryly observes, marked by much greater attention to her physical appearance:

> "Did you ever wonder why lady comrades are such frumps?"
>
> "I never noticed," David said.
>
> "If you're writing a play," Abbe said, "and you write a Party woman into it, you get a woman with stringy hair, bad skin, thick glasses, and uneven hems. That's cartoon thinking, but it's sometimes accurate....There's a reason for this. In the Party, ill-fitting clothes, sloppy make-up, these are proof that you accept that part of the code that says that sex is of no special importance and that romantic attachments are bourgeois impediments. Sex is like a drink of water. . . . On Fourteenth Street, they used to say that before Comrade Klein went out on the street, she stood in front of a mirror and looked back over her shoulder to make sure her seams were crooked."[11]

Abbe's concern for David—who she fears is "going to get hurt, chopped to pieces for a principle you didn't know existed"—and, presumably, her recovery of bourgeois values redeems her: after she exposes Barney Castle's perfidy she disappears from the novel.

If Abbe Klein represents a Communist *femme fatale*—albeit one with redemptive possibilities—Angel Chavez's mother, Consuela, is transformed into a parody of a "Scottsboro Mother." Consuela Chavez is initially depicted in sympathetic, if somewhat stereotypical, terms: "hers was a classical Mexican face, thick, sturdy, the skin somewhat darker than her son's."[12] During a critical moment early in the novel, when a lynch mob gathers outside of the San Juno jail, Barney demands Consuela's complete trust, and she accedes to him. From this point on, the novel makes clear, she is hopelessly ensnarled and corrupted by Barney's manipulation. One early sign of her seduction by left-wing political culture appears in the form of a press notice:

> "'*Consuela Chavez*, whose boy will be tried out there for murderrape' the item ran, 'will script her life-yarn for *Struggle*, a race organ which is having

one to break even. Heavy sugar paid her will go to the boy's defense fund, itsezzere'. . . ."

"But she can't write," David protested. "Can she?"

"She can endorse the checks," Abbe said. "She's literate."

"That's not what I meant," David said. . . . "Can she write her life story? Write it so people will want to read it?"

"Of course not," Abbe said sharply. "Neither can the ball players who write the inside story of the World Series. She's a celebrity. Celebrities don't write their life stories. They tell them to bright young men with horn-rimmed glasses, and the bright young men turn in the stories, and bright old men fix them up. What's the matter with that?"[13]

Another step towards Consuela's transformation into a political celebrity is revealed in the following exchange between Barney and one of his associates, Cheever, over a script that has been written by Party hacks for her to broadcast:

"This is pure crap," Barney announced at last. "I don't object to party-line stuff, but this business about"—he read scornfully—"'to kill my boy is but a way my country make war on her dark-skin peoples' is plain nonsense, and everybody that hears it is going to know its nonsense."

"It's been cleared," Cheever said.

"Not by me." Barney's eyes roamed the page he was holding, and then he read again: "'Even if my boy die, he no die for nothing.' Why the mammy dialect. Consuela's English is all right."

"My speech is awkward," Consuela said diffidently. "But not that bad."

"I was trying," Cheever explained, "to give it the Sacco touch."

Barney's laugh was loud and ugly. "Christ Almighty!" he shouted. "You guys throw some soap-opera dialogue on top of some used up *Worker* editorials and talk about Sacco."[14]

While Consuela appears as an object of Party manipulation in this passage, later in the same scene she reveals her capacity to negotiate on her own behalf:

"But will there first be a dinner?"

"Yes," Barney said. "You and me and the rally committee. But I'll get you out of it. It's customary, that's all."

"They work for Angel," Consuela replied. "If they wish to see me, I am honored. But I wonder—would it be too much to ask—"

"Go ahead. If it's possible—"

"I'm ashamed." Consuela turned to David. "You will think I am ungrateful." Beneath her dark skin, David realized with surprise, Consuela Chavez was blushing.

"Name it," Barney urged. "You've done all I asked you for."

"Not frijoles," Consuela blurted out. "Not the corn that comes from cans that they call Mexican corn. I never in my life saw it before I came here. No tamales. No enchiladas. One does not eat these for dinner....At home we eat meat—beef, pork, sometimes veal—and potatoes and vegetables, and sometimes canned pork and beans. But my diet here would give ulcers to a peon."

Barney laughed. "You hungry now?" Consuela nodded. "How about a steak?"

"Now?"

"Right now." Barney picked up the phone, called room service and ordered, pausing after each item to be assured by Consuela's nod that he was choosing well, a shrimp cocktail, filet mignon, medium rare, baked potato, apple pie, and coffee.[15]

This comic episode prefigures the deeper transformation Consuela undergoes by the end of the novel. Composed and self-assured when she addresses an audience of thousands at a New York City rally for Angel, Consuela delivers a comparable public performance at the moment her son is declared guilty:

From the defense table, dodging across the aisle, Consuela Chavez ran, her black hair unkempt, falling across her face. Through the crowd she came, screaming her boy's name as she ran. Judge Motley, who had been about to enter his chambers, turned. She saw him and ran to the space before the bench and looked up at him and shook her fists.

"You kill my boy!" she screamed. "Santa Maria! You kill my boy, my Angel!"

The photographers swarmed around her, some climbing on the witness stand, pushing the witness chair down in order to get better angles, and their flashbulbs popped as Consuela continued her disorderly harangue. What made the moment uniquely horrible was that David could see that, even as she gestured toward the Judge, she turned her body so that the corner of the Judge's bench did not hide her from the lenses, and that her gestures became more pronounced and her face more distorted whenever a flashbulb popped.[16]

Mankiewicz reserves his sharpest attacks for the figure of Barney Castle, the crafty, behind-the-scenes operator whose actions are responsible for Angel Chavez's death. Barney is absent for a great deal of the action of *Trial*, in accordance with a division of labor that makes David Blake responsible for the day-to-day conduct of the trial while Barney assumes the responsibility for fund-raising. This division of labor allows Mankiewicz to explore in some detail the behind-the-scenes operations of "popular front" politics. When Barney insists that David fly to New York City for a weekend to participate in a fund-raising rally for Angel, David is introduced to the byzantine world of left-wing finances. As Barney tells him: "In the next few hours, you are going to meet so many people, and hear about so many organizations, it'll have you as dizzy as its got me. I'll try to explain what I can, but you'll just have to take my word for the rest."[17]

David's introduction to the wheeling-and-dealing of left-wing politics sets the stage for the dramatic disclosure, at the end of the novel, that the Party raised well over three hundred thousands dollars for Angel Chavez's defense—two hundred thousand of which was diverted to other Party purposes. This revelation clearly echoes the long-standing accusation against the ILD and the Communist Party that they had freely used the substantial funds collected for the defense of the Scottsboro Boys for other purposes.

And there is more to Barney's treachery. At the critical moment in the trial, as David prepares to make his closing argument without calling Angel to testify on his own behalf, Barney reappears in the courtroom, requests a recess, and bullies David into putting Angel on the witness stand. Bewildered, David reluctantly agrees. Abbe understands Barney's tactics all too well; after he leaves she explains:

> "If, by some miracle, Armstrong [the prosecutor] screws up and doesn't make a monkey out of Angel, Barney'll think up some other way to blow the case."
>
> "Why, Abbe?" David asked. "Why do you say that? All right, maybe Barney's got other interests; maybe he doesn't care if Angel lives or dies. But why do you say he wants him guilty? It's bad enough to say he doesn't care. . . ."
>
> "He does care," she said. "He cares terribly. He wants to see Angel hang."[18]

Angel is indeed executed, and thus does *Trial* embody the second charge hurled against the Communists in their handling of the Scottsboro case: that they would rather see the boys dead than alive, that they secretly hoped for—and perhaps worked for—their execution as the best way to expose the bankruptcy of the American judicial system.

Mankiewicz could have easily concluded *Trial* with this anti-communist diatribe, but he clearly believes that the Communists constitute only one dimension of the threat to the American republic. The other danger appears in the form of a committee that sounds remarkably like the House Un-American Activities Committee: the State Assembly Committee on Disloyalty and Subversion, chaired by Carl Baron Battle, its first and only chair. Mankiewicz leaves no doubt about how Battle should be viewed:

> He regarded Carl Baron Battle as an evil man, a subverter of the democratic institutions of the State, and, worse, as an opportunist who played the part of a fanatic because, in an age of fanaticism, extremism offered at-

tractive possibilities, mostly financial. He knew there were others in the State who regarded Battle as a savior, as the only man who could save the State (indeed, perhaps, the nation) from the alien "forces" which threatened to engulf it.[19]

Battle relentlessly pursues Blake with one subpoena after another during the course of the Chavez trial, and David continues to resist appearing before his committee — in spite of pointed warnings from Abbe and an apparently chance encounter with Charles White, the publisher of the oldest newspaper in the state, who warns him during a plane flight to New York: "you are heading for bad trouble. Very, very bad trouble. Just about the worst trouble a man, in this day and age, can get into" — adding, significantly, "you're getting plenty of help. . . . You have plenty of help getting into it."[20]

In spite of these increasingly dire warnings, David seems to believe that his honesty and personal integrity will prevent the Committee from destroying his life and reputation.

When — after the failure of his appeal to the State Supreme Court, after the execution of Angel Chavez, after an acrimonious newspaper exchange with Carl Battle — David finally appears before the Battle Committee, he is devastated by the deception Barney and the Party has practiced upon him. His appearance is preceded by that of Epstein, the treasurer of the Free Angel Rally, Provisional Committee and a former Party member, who offers in excruciating detail the web that ensnared David:

> "He [Barney] said he'd heard of a case that might interest us, and we wanted to know if he should intervene. . . ."
>
> "Well he told me about the case. Said it met all our requirements. The Mexican boy, the sex angle. It was ideal for us, he said. He even said there was a continuing interest."
>
> "Hold it," Battle commanded. "What did he mean by that?"
>
> "He meant that we'd have something left after the case was over. Something that would justify all the expense and trouble we'd have to go to to handle the case."
>
> "Did he tell you what the continuing interest was?"
>
> "Yes. He said the boy's mother could be used as a rallying point for years after her son was executed. . . ."
>
> "I said we'd consider it, if he got a local man to do the actual courtroom work, someone who hadn't been associated with this sort of thing before."
>
> "In other words," Battle suggested, "a front."[21]

David is further shattered by the behavior of his mentor, Terry Bliss, who tracks him down to announce that he has — without David's permission — requested an unpaid leave of absence from the university for him, so that the university will not be compromised by his appearance before the Committee.

Furthermore, terrified by the threat of guilt-by-association, Bliss insists that David refuse to testify by invoking his Fifth Amendment privilege. Hemmed in on all sides, David is plunged into suicidal despair. At the last minute, however, a *deus ex machina* intervenes in the form of the newspaper publisher, Charles White, who persuades David to draw upon his private journal and write a full account of all the details of the Chavez case, which he will run in installments in the *State Record*. As he patiently explains:

> "Until you write this—until I print it—the truth, as you know it, does not exist, except for you. As far as Battle is concerned, you're a blank slate, and as far as the rest of the world is concerned, they either don't care or they accept what Battle writes on you—on the slate—as truth. That's what Battle likes, to put his lies up against nothing, against ignorance. But they won't stand up against the truth."[22]

Apparently, a free unfettered press is the best defense against the twin threats of Communism and the assault on civil liberties.

Although *Trial* was published to decidedly mixed reviews, and attacked by some critics for its wooden characters and implausible coincidences, perceptive critics immediately grasped its cinematic possibilities. Riley Hughes, writing for the *Catholic World*, commented: "*Trial* reads like the slick movie scenario (after bowdlerizing of the sex scenes) it will undoubtedly become; it is dismaying to think that this smooth, superficial hokum should be considered worth the $10,000.00 prize it received."[23] The (London) *Times Literary Supplement* also predicted a Hollywood future: "The author adheres closely to his plot, but it is only possible to believe in his people by visualizing what actors will give them the illusion of life when this literary property reaches its predestined terminus, the silver or 3D screen."[24]

Trial, with a script by Don Mankiewicz in 1955, did indeed fulfill its destiny as a film, coming at the tail end of the spate of more than fifty anti-communist films produced by Hollywood between 1947 and 1954.[25] In director Mark Robson's capable hands, an excellent cast—including Glenn Ford, Dorothy Maguire, Anthony Kennedy, Juano Hernandez, Rafael Campos, and Kathy Jurado—dramatically heightened the script's anti-communist outlook. Glenn Ford, who specialized in playing ordinary, unprepossessing characters who somehow always manage to summon up extraordinary courage under pressure, plays David Blake with an increasing air of frustration and aggrieved innocence as the film proceeds. Nora Sayres has wryly and astutely noted that anti-communist films "instruct us especially on how American Communists look. . . .you can sometimes spot a Communist because his shadow looms larger and blacker than his adversary's. Also, movie-Communists walk on a forward slant revealing their dedication to the cause."[26] This characterization is certainly true of Arthur Kennedy's portrayal of Barney Castle; in addition, his lean, angular face is contorted into a sneer throughout the film. Sayres also notes that there is a type of woman in this genre that she identifies as the "Bad Blonde": "in the Fifties, you knew that there was something terribly wrong with a woman if her slip straps showed through her blouse; in this context, it meant treason. Bad blondes tend to order triple bourbons or to be hooked on absinthe, and they often seduce 'impressionable' young men into joining the Party."[27] Dorothy Maguire plays Abbe; her slip straps do not show, but viewers do get a glimpse of her bra in an early scene with Glenn Ford. Juano Hernandez, whose breakthrough moment had occurred with his dignified portrayal of a black man unjustly accused of murder in the 1949 film *Intruder in the Dust*, brings a comparable sense of dignity and gravitas to his portrayal of Judge Theodore Motley, while Rafael Campos—who often was cast as a troubled Hispanic youth, as he was in *The Blackboard Jungle* (also with Glenn Ford)—effectively conveys Angel Chavez's innocence and confusion; and Kathy Jurado brings equal measures of concern and opportunism to her portrayal of Angel's mother.

In the film version of *Trial*, the threat represented by the Battle Commission recedes, displaced by the film's emphasis upon Communist manipulation and racial harmony—and the ways in which Communist intrigue poisons and undermines healthy race relations in the United States. Early in Angel's trial, David Blake—spurred on by Barney's coaching—has a confrontation with Judge Motley in his chambers:

> "Angel is not going to get a fair trial, only the appearance of one. I don't like the whole thing. . . . And I'm going to tell you something else. I don't like you presiding. I think you were selected so that after the conviction they'll be able to say: 'Of course the Mexican had a fair trial; sure, he even had a Negro judge. . . .'"

In the face of Judge Motley's stern but silent rebuke, Blake fumbles his way back to his senses and comes to an embarrassed silence. The film makes it clear, however, that Blake has gone too far and that, indeed, he is functioning as Barney's mouthpiece in this scene. The first sign that David has recovered his moral compass occurs when he attends the rally of the All People's Party in New York, rips up the speech Barney has written for him, and musters up all of his courage to challenge Communist deceit: "I've seen and heard a lot of things tonight that I don't . . ."—only to be deliberately drowned out by the blare of band music. As he later tells Abbe:

> "I ended up on the stage with Communists out there in front of me, Communist banners in back of me. I tried to tell them what I thought of them and I ended up just talking to a dead microphone with a brass band blaring. . . .
>
> "Five minutes after I got to New York I knew that Barney was a Communist. Five minutes after I got on the plane to come home I knew that if Barney was a Communist you must be one, too."

Abbe remorsefully confesses: "The Party's collected money for a martyr; now it must produce one." In rapid fashion, Barney Castle forces Blake off the case; Angel Chavez is found guilty, to the obvious elation of the jailor, Fats Sanders, who has earlier in the movie indicated his social and political affinities with the mob that has gathered outside of Chavez's cell—promising it a "legal lynching" if it restrains itself; and Mrs. Chavez self-righteously rejects Blake's offers of further assistance: "If my son must hang, at least he will hang for something worth hanging for. He will be mourned by millions of people all over the world."

Yet, in a series of sudden reversals, Mankiewicz's film script dramatically undercuts the conclusion of the novel. Appearing at Angel's sentencing as a friend of the court, in spite of Barney Castle's vehement objections, Blake makes a striking confession:

> "I was a fool and this boy suffered because of my foolishness. . . . Bernard Castle and the Communist Party had decided that Angel Chavez should die, so that his death might stir up more race hatred, raise more money for the Communist cause . . . In my anxiety to help this boy, I permitted myself to be used as a Communist front."

Apoplectic with rage, Castle turns the full force of his fury upon Judge Motley:

> "You ought to disqualify yourself for incompetence. You're a frightened little man, selling out his own people for a fancy title and a black gown. They may call you 'judge' to your face, but they got a better word for you behind your back. You're a handkerchief head, you're an Uncle Tom. . . ."

Castle's rabid outburst is so savage, so racist, so patently unfair, that even apparent racists like Fats Sanders are stunned. The entire courtroom unites in a gesture of human solidarity around Judge Motley. In one of many dramatic conversions that occur in the final scenes of the film, Blake solemnly announces:

> "There was a time when I believed that this boy was being railroaded to his death because of his race. But somehow I believe that all of us here have learned an awful lot here these last few weeks. . . . And now I think that these people, just like myself, they ask only one thing for Angel Chavez, and that is justice. . . ."

Chastened by Blake's speech, the District Attorney (played by John Hodiak) joins in his appeal for fair play for Angel. The jury and the courtroom applaud them and Judge Motley, appropriately moved, sentences Angel not to death, but to an indeterminate sentence at the state industrial school. All is forgiven as the entire courtroom joins in celebration of bedrock American values—except, of course, the nefarious Barney Castle, who is given a thirty-day jail sentence for contempt of court. And Angel's mother regains her senses, admitting to Blake that she was wrong in her embrace of Castle and the All People's Party. Almost as an afterthought, the subpoena server for the Battle Commission informs Blake that Battle has postponed his scheduled appearance—indefinitely. All's well that end's well.

If *Trial* represents one of the ways in which the Scottsboro case was refashioned by the exigencies of Cold War America, Grace Lumpkin's recantation of her previous political views is an even more striking transformation. In retrospect it seems that Lumpkin's political radicalism had always existed in uneasy tension with her family heritage and social upbringing.[28] In between the critical successes of *To Make My Bread* in 1932 and *A Sign for Cain* in 1935, Lumpkin—under the pseudonym Ann DuPre—published two novels concerned with social milieus far removed from those she had explored in her other fiction. *Some Take a Lover* and *Timid Woman*, both published in 1933, place at their center young women who are struggling against the constraints of the values of their families. Set in New York City, *Some Take a Lover* recounts the stories of the children and grandchildren of a tyrannical matriarch, Mrs. Singleton, "who watches with cynical interest their extra-marital intrigues, while holding over them the threat of disinheritance in her will. Only one of the group retains her grandmother's respect, Natalie, who 'lives in sin' with a young artist but is loyal and faithful."[29]

More intricately crafted, *Timid Woman* opens on the eve of the party that will announce the eligibility for courtship and marriage of eighteen-year-old Ellen Guaillard. The product of a respectable and conservative southern family, Ellen chooses instead to remain single and live at home, devoting herself to the care of her family. When her family's financial decline forces her to seek work elsewhere, Ellen—at the age of thirty—takes a position as governess for the children of Alex Kingman, whose wife is confined in a sanitarium. Deeply concerned about appearances and gossip about Ellen living in the same house as a married man, her family exhorts her to return home. In the denouement of the novel, however, Ellen refuses to accede to her family's wishes, thus signaling her triumph over its conservative values.

The parallels between Ellen, whose lack of money in a once more affluent family propels her to seek financial and social independence in the larger world, and Grace Lumpkin herself are striking; indeed, Lumpkin habitually drew upon personal and family history as the basis for all of her fiction. In this regard her novels function as a register of her march from radical politics to the intensely conservative political outlook that characterized the end of her career. One of the casualties of this journey was the Scottsboro case.

During her years on the Lower East Side in the early 1930s, Grace Lumpkin and Michael Intrator had become lovers and may have married; Lumpkin's roommate, Esther Shemitz, and Whittaker Chambers married in 1931 and, for a while, cohabited with Lumpkin and Intrator. By then, Intrator—described as "brash, confident, and irreverent toward the Communist Party"[30]—had been expelled from the Party. Lumpkin, by all accounts, remained in the Party's good graces until the late 1930s. In the meantime, Whittaker Chambers became increasingly drawn into the Communist Party underground. In 1932 he and Esther Shemitz left New York and, for the next six years, lived furtive lives, assuming different pseudonyms and moving from place to place.

Lumpkin and Intrator remained in New York. Intrator was completely ostracized by the Communist Party, but Lumpkin, apparently, sought to balance her loyalties among different factions. According to Suzanne Sowinska, however, the greatest crisis Lumpkin faced, sometime between the mid- and late 1930s, was not political; it was deeply personal: she became pregnant with Intrator's child and decided to have an abortion. Shortly after this experience, her relationship with Intrator came to an end.[31] Lumpkin had come to a major crossroads in her life, one captured in the deceptively simple surfaces of the third novel published under her real name, *The Wedding*.

Set in the fictional town of Lexington (drawing upon Lumpkin's memories of her childhood in Columbia, South Carolina), *The Wedding* revolves around the multiple domestic crises that break out on the day of Jennie Middleton's wedding in 1909. The daughter of a socially prominent but financially strapped family, Jennie is engaged to Dr. Shelley Gregg, the product of a frugal, hard-working mountain family. The main action of the novel is triggered off the evening before the wedding, when Jenny and Dr. Gregg have a dispute that quickly escalates into a shouting match. On the morning of the wedding Jennie announces to her parents that she will not proceed with the wedding.

Jennie's father, Robert, tries to mediate the dispute, but he is clearly troubled by the financial burdens of the wedding. There are also serious tensions between Robert Middleton and Dr. Grant, the new "Yankee" minister of the family church. Mr. Middleton has effectively barred Dr. Grant from officiating at Jennie's wedding and Dr. Grant has indicated his disapproval of the use of Confederate memorabilia during the ceremony.

Meanwhile, Jennie's younger sister, Susan, is involved in a drama of her own. As a child in the Middleton household, Susan has immediate and intimate access to the African Americans who keep it running smoothly, Louisa and Ed; she is also intimate with Old Rosin, a white mattress repairer now serving a life sentence in prison for killing a man when she was drunk. Susan foolishly allows her wealthy, spoiled cousin—ironically named Saint John Middleton—to enlist her in acting out a memorable episode of family history during the Civil War: hiding the family silver from the Yankees. In this instance, the family silver is represented by one of the wedding presents, a silver bowl—which Saint John steals and orders Ed, the black servant, to bury after

Saint John has wrapped it in Ed's coat. When the theft of the bowl is discovered, Saint John publicly accuses Ed and Susan's silence seals Ed's fate; he flees the house in fear of apprehension by the police.

Other crises arise in *The Wedding*, but the respective rebellions of Jennie and Susan against the constraints imposed upon them as southern women constitute the moral centers of the novel. Although contemporary reviewers received the novel as a comedy of manners, southern-style, in the words of one of them "a slight simple story as far from the subject of her first book as a picket line is from a pulpit,"[32] Lumpkin's novel quite effectively probes the limits of her characters' consciousnesses—raising some tantalizing hints about her own political straddling act.

Jennie's spirited, last-ditch rebellion against her impending fate as a married, bourgeois southern woman is the expression of what Jessica Kimble Printz calls a "protofeminist" consciousness[33]—so described because, in Kimble's view, "Jennie is acutely aware of her own experience of sexual oppression, but she never translates this personal experience or the glimmering of feminist reflections it inspires into an appreciation of the 'predicament' of other women, not even women in a similar social position. Born in isolation, her feminist impulses remain stunted and ultimately ineffectual."[34]

Similarly, Susan's revolutionary sympathies for the situation of African Americans and white workers compel her to break out of her guilty silence and speak out forcefully on behalf of Ed—literally slapping down her cousin, Saint John, in the process. Although she manages to convince her father of Ed's innocence and Saint John's guilt, the story she tells has little real impact on the dominant attitudes or the claustrophobic social structure of the world of the novel: Ed remains banished from the household, Saint John goes unpunished. Like Jennie, who accedes to her predetermined role at the end of *The Wedding*, Susan, too, retreats—in her case into the realm of religious imagery to try to make sense of the injustices she has witnessed:

> Suddenly a terrible sadness came in Susan. . . . She thought of Ed running in some place that was like a desert. And he was alone. He was running like the song that Louisa sang, "Keep arunnin', Keep arunnin', the fire gwine overtake you." She saw Ed running with fire trying to overtake him, a fire in a cloud, like the fire that came down with the ark of the Israelites in the Bible. The fire moved forward swiftly over the desert-like place so that Ed must keep running to escape it. She saw the great ball of fire enclosed in a cloud and a dark form, frightened and alone, hurrying before it.[35]

Although the novel is couched in the language of domestic melodrama, the analogy between this narrative of willful racial injustice and the plight of the Scottsboro Boys is too striking to be overlooked—particularly in light of Lumpkin's role as one of the most prominent literary radicals of the 1930s. At the same time, the subsidiary position of this event within the dramatic structure of *The Wedding*, combined with Lumpkin's intricate and clinical exploration of the limitations of Jennie's and Susan's consciousness, strongly suggests a deep crisis in Lumpkin's earlier revolutionary fervor.

The publication of *The Wedding* roughly coincided with Lumpkin's abandonment of the Communist Party and her active involvement with the Moral Re-Armament Movement (MRA). Initially founded by the American evangelist Frank Nathan Daniel Buchman in the 1920s as an international movement calling for moral and spiritual renewal, the MRA was first known as the Oxford Group; in 1938 it changed its name to Moral Re-Armament and became more directly involved in political and social issues, adopting a sharply anti-communist outlook that deepened, and found a receptive popular and political climate, during the Cold War. Devout practice of the "Four Absolutes (honesty, purity, unselfishness, and love)," the MRA insisted, was the only way to achieve social, political, and economic justice.

The national headquarters of the MRA was Calvary Church in New York City, presided over by the Reverend Samuel Shoemaker, a prominent Episcopalian minister. One of the practices of the MRA, "sharing," encouraged recent members to confess their public and private sins, either privately to another MRA member or clergyman or publicly at group meetings or house gatherings; in this context, Reverend Shoemaker soon emerged as Lumpkin's father-confessor and a major influence upon her life.[36] When, in 1941, Shoemaker called for a return to more orthodox religious practices and split from the MRA, expelling it from Calvary Church, Lumpkin aligned herself with him and rejoined the Episcopal Church. She eventually moved into the parish house of Calvary Church, quietly living there until the late 1940s, when she was drawn into the vortex of the Alger Hiss case.

Lumpkin's old comrade Whittaker Chambers had resurfaced in the late 1930s, now a staff member of *Time* magazine and a virulent anti-communist. He wrote a spate of articles attacking both Communists and New Deal liberals and gave testimony before various government agencies, detailing his experiences as both an open and covert member of the Communist Party. Initially ignored in the early years of the Cold War, Chambers was rediscovered by politicians increasingly preoccupied with the Communist threat—particularly after 1948, when he testified before the House Committee on Un-American Activities (HUAC) that he had been involved with Alger Hiss as part of a Communist Party underground network when Hiss had been a State Department official.

After Hiss's indictment for perjury in 1949, Lumpkin was visited by a private investigator hired by Hiss's defense, seeking information about Chamber's "deviancy" that would destroy his credibility on the witness stand.[37] Instead, Lumpkin emerged as one of Chambers's most loyal defenders. As the Alger Hiss case came to dominate American political life and popular consciousness, Lumpkin emerged out of her period of relative seclusion to make common cause with Chambers's anti-communist crusade. For the next several years, Lumpkin lectured on the dangers of Communism at churches throughout New England. After the publication of Chambers's book *Witness*,[38] which she proclaimed as "one of the great books of the 20th century,"[39] Lumpkin sought to link herself even more closely with Chambers—claiming that it was Chambers who had been responsible for her abandoning Communism in 1938.[40]

In 1953 Lumpkin happily appeared before the Senate Sub-Committee on Government Operations, freely giving testimony against her former associates and purging herself of the taint of her Communist past. Shortly after this appearance, Lumpkin left New York City for good, spending several years in Virginia before returning home to Columbia, South Carolina. Until her death in 1980, she spent the rest of her life speaking and writing about the evils of Communism from a born-again Christian perspective.

It is against this backdrop that Lumpkin's last published novel, *Full Circle* (1962), can be understood—the novel that not only reenacts her political apostasy, but also dramatizes the ways in which she sacrificed the Scottsboro Boys to the exigencies of Cold War politics. Like Lumpkin's earlier works, *Full Circle* freely draws upon actual historical figures and events, in this case Mrs. Craik Speed and her daughter Jane, who were much praised by the Communist Party for their activities during the early years of the Scottsboro case. Mrs. Speed was the granddaughter of the founder of Montgomery, Alabama's First National Bank, and a member of the Daughters of the American Revolution and the Colonial Dames. Her ancestors included some of Alabama's most distinguished doctors, lawyers, jurists, and planters. She was also a dedicated follower of the Socialist Norman Thomas and was, therefore, regarded as somewhat eccentric. Her social status provided her with somewhat of a cushion, however, and the citizens of Montgomery tolerated her eccentricities. In 1931, Mrs. Speed became deeply interested in the Scottsboro case and, encouraged by Jane, then twenty-one years old, she began to read all she could about it. At first she was confident that the "good citizens" of Alabama would see that the "Boys" received just treatment; by 1933, however, she and her daughter had openly endorsed the efforts of the International Labor Defense to free them. When Jane addressed a May Day "united front" rally in Birmingham, she was arrested and charged on four counts: "disorderly conduct, speaking without a permit on a public thoroughfare, and two counts dealing with speaking to a public assembly of mixed races without physical barriers of separation."[41] Rather than pay the fine, Jane Speed insisted on going to jail. In Montgomery, police placed Mrs. Speed's home under surveillance and she was completely ostracized by her friends. When Jane was released after serving fifty-three days in jail, she and her mother left Alabama and moved, in Mrs. Speed's words, into exile (Rabbi Benjamin Goldstein, who had also been involved in Scottsboro activities in Montgomery, and who had testified as a character witness at Jane's trial, was forced to leave the state around the same time).[42]

Full Circle fictionalizes the Speeds' experiences. Somewhat like *The Wedding*, *Full Circle* is presented as a series of domestic crises. Narrated in the first person by Anne Braxton, the novel begins with an account of her daughter Arnie's nervous breakdown in New York City. Mrs. Braxton brings Arnie home to the South in the hope of restoring her physical and mental health. Through a series of flashbacks, Anne Braxton's life slowly unfolds: for the past twelve years she has been estranged from her twin sister Julie, who still lives in the family home in the small southern town of Bethel; she is divorced from her husband David, the former rector of their church in Bethel, who has begun to question the tenets of his religious faith and has fallen in love with a younger woman; since her divorce she has been actively involved with the Communist Party, a commitment initially driven by Arnie's emotional response to a Scottsboro-like trial in their hometown:

> She had seen people going into the courthouse; and she had followed them inside to the room where the trial of two Negro boys was being conducted. We had, of course, known about this trial for some time. The two boys had been arrested in April and had been in prison since. At first their arrest had been reported only in our local newspaper. Then, as the date for the trial came nearer, Northern papers had made the case headline news.
>
> The events that led to the arrests of the two Negro boys were simple. One day in April they had met two white girls on a freight train on which all were stealing a ride. The girls (they were eighteen and twenty-one) were acknowledged prostitutes, though the younger, Posie said to me later, which [sic] I met her in New York while she was under the protection of the Party, that the older girl, Lilly, had persuaded her into the life she had been living.
>
> I believe no one has ever been able to distinguish the truth amidst the sordid details given, and the terror and hate and bitterness which the trial aroused in our town and in our part of the country.[43]

The basic facts of the Scottsboro case are radically reconstituted through the words and perspective of Mrs. Braxton, setting the stage for the displacement of the case in the world of the novel. At first it seems as if Mrs. Braxton's sympathies rest firmly with the two black defendants, even if her feelings are refracted through the lens of her obvious class and social biases:

> To me the immediate blame for all that happened should be put upon the sheriff and the other men who took the four young people from the train, accused the boys of raping the girls, and arrested them. Knowing well the reputation of the girls, they must surely have understood that women of that sort made no discrimination as to color, any more than had many

white men of the South, or the Carpetbaggers and soldiers who came from the North and occupied the South during Reconstruction, when they desired someone of the opposite race.[44]

But the convoluted logic of Mrs. Braxton's subsequent observations makes it plain that she continues to think and function as an unreconstructed daughter of the Confederacy:

But though I put the immediate blame on the sheriff, nevertheless, since I am trying to be factual and objective, it is necessary to realize that there was much truth in the often ignored facts of Reconstruction, which made even the second generation after that time fearful of control by Negroes and Carpetbaggers—just as the Jewish refugee who tuned my piano found it difficult to believe that the Nazi regime would not have a resurgence of power in Europe.

This fear or dread was expressed by the sheriff one day during the trial when I had gone with Arnie to the courthouse. Coming out of the courtroom after the noon recess had been declared we met the sheriff in the corridor. There he angrily berated Arnie for her stand about the Negro boys. People were still leaving the courtroom and they gathered about us as the sheriff, a large man with a powerful voice, in vivid language reminded Arnie of the Southern women who had been raped by Negroes during Reconstruction. . . . Arnie was furious, but I, having heard my father speak of those days . . . knew that there was some truth to what the sheriff had said.[45]

Through Anne Braxton, assuming the posture of objective detachment, *Full Circle* echoes some of the charges hurled against the Scottsboro defenders by southerners:

It had been rumored early that Communists had come into the community to stir up trouble over the case, but I paid little attention to the rumor, knowing that each side was making unwarranted statements about the other, and that a Northern lawyer had been brought down to defend the boys—which was an added irritation to the Southern whites, even those who were originally in favor of writing the trial off as a mistake of the sheriff and his deputies, and letting the Negro boys go free.[46]

But Arnie is swept up in the passions of the case, and her mother accompanies her:

People who remember the case will also remember the eighteen-year-old girl, her brown eyes blazing, who, at meetings held at night on the public square and in the Negro district and in the poorer sections of town, stood on an improvised platform and asserted the innocence of the two Negro boys to the crowds who gathered about her. That child was Arnie.[47]

In this way the force of the 1930s Scottsboro Narrative is undercut in *Full Circle*. At any rate, Scottsboro is not central to the plot of the novel; it is only the springboard for Lumpkin's exposé of the evils of Communism. From this point onwards, Arnie and her mother are increasingly drawn into the byzantine world of Communist Party politics.

As "young men from New York" stream into town to cover the case, Anne opens her house to them; slowly, she is drawn into the orbit of revolutionary discourse as her bourgeois illusions are exposed to criticism and ridicule by them, and broken down. After an intense period of discussion and ideological indoctrination, Anne, following her daughter, applies for membership in the Communist Party. Still, tensions and contradictions from her former life continue to plague her. In one telling episode, on Christmas Eve of 1937, Anne and her twin sister Julie resume their long-established family practice of decorating a Christmas tree and setting up a crèche—a task that used to belong to Arnie before she became a Communist. When Arnie and her lover Art return home that evening, Art commits an act of sacrilege against the crèche by symbolically lynching the figure of the Christ Child and overturning the figures of Mary and Joseph.

Outraged, Julie unleashes a stream of invective at Anne, Arnie, and Art:

> "'I thought you were clean, Anne. I thought your daughter was clean, but both of you have become filthy, no better than those girls who enticed two ignorant Negro boys. Do you hear? You are no better than they. . . .'"[48]

When Anne challenges Julie to rise above her complacency and recognize that she and her comrades have risen to the challenge of "fighting the terrible scourge that is in Europe, the shocking persecutions, the tyranny, getting ready to destroy us all," Julie rejoins: "I hope before God they will come to this country and that they will destroy you and your whore and your Jews. I hope to God they do."[49]

Art, who is Jewish, then inveighs against Julie—in the process unwittingly supplying ammunition for widespread theories and beliefs about Jewish conspiracies for world domination:

> "So that's your good will and love. You talk about the Jews," he said. "In another twenty years those Jews you despise so will be ideological rulers of this country. The Jews, the only progressive people in the world will have influenced every mind. . . .The Jews. . . ."
>
> "Hush, Art," Arnie said. "Hush!" But he did not heed her. "You talk about filth," he stammered. "We are the only ethical people in the world. Why the Jews you despise so can drag a Goy into the gutter and never even get a little finger stained with filth. Those Jews you despise so. . . ."[50]

This scene is the basis for the rupture between Anne and her sister, for she unconditionally defends Arnie and Art, even when she has serious private reservations about their views. As in the case of the trial of the two African Americans, Anne's "objectivity," reinforced by her ferocious pride in her family heritage, permits her to make allowances for her sister's vicious anti-Semitism:

> Was Arnie right and should I accept the words Art had said lightly as if "they did not mean a thing"? Yet there was my grandfather's diary. During the Civil War he had lived in Richmond, a member of the Confederate Headquarters there. Scattered through his diary from the beginning to the end of the war, on different pages, there was written down the animosity felt by all toward Judah P. Benjamin because of his influence over Jefferson Davis. And while I knew that some of the animosity must have come from prejudice and jealousy, there was still what had seemed a "legitimate" bitterness because Benjamin had ridiculed General Lee's belief in God, and had exercised his caustic wit on Stonewall Jackson whose chief fault, in Benjamin's eyes, was that, before a battle, the general knelt down and prayed. Was there, then, something that cynically and consciously defiled?[51]

Arnie's illness, we learn, has been triggered by her expulsion from the Communist Party for racial prejudice, an accusation made against her during one of the Party's periodic campaigns against "white chauvinism"—shades of the 1931 trial of August Yokinen. The accusation has been made by Arnie's husband, Ben, and it extends to both Arnie and her mother. Struggles for racial justice had been central to Lumpkin's fiction from the beginning, but *Full Circle* calls into question the very integrity of her earlier works through Anne Braxton's stinging, thoroughgoing critique of the callousness, the cynicism, the manipulation, and the fundamental inhumanity of Party members. As she methodically recalls every detail of what she now regards as her ideological indoctrination, the trajectory of the novel moves relentlessly towards the South, towards another form of conversion.

One clear sign of Lumpkin's ideological apostasy is her representation of African American characters. In sharp contrast to the figure of Selah, the young black woman who emerges at the end of Lumpkin's second novel, *A Sign for Cain*, she offers in *Full Circle* the character of Mum Hattie, the loyal black nurse and family retainer who is always there during moments of crisis. Mum Hattie, whose story is also told in flashbacks, has previously left the South to pursue a nursing career in New York City. She reappears in the South with an irresponsible husband who is later killed in a brawl in Harlem, and leaves again, traveling back and forth between New York and Africa, where she has established homes for orphaned children and babies. Mum Hattie functions as a source of stability for Anne and the voice of faith and healing in *Full Circle*—the clear alternative to the Machiavellian world of the Communist Party.

After informing Anne that her sister Julie has reappeared and plans to seek psychiatric care for Arnie and place her in a sanitarium, Mum Hattie explodes into a diatribe about Julie, now an observer for the Covenant on Human Rights of the United Nations:

> "Human Rights, social this and social that! Where do these human rights come from—where does that social security everybody talks about come from? Where does that equality come from? Not from that Tower of Babel on the East River in New York. Here in this country there's a heritage, and we sold it at San Francisco for a mess of pottage.
>
> Who gave anybody rights? Did Thomas Jefferson? Did Lincoln? Did Lenin? My rights, gimme my rights, everybody says. But it's what your ma told us: 'Rights are from God. Rights have got to be earned. . . .'"[52]

If Mum Hattie functions as one important guide for Anne's return to her southern heritage and Christian faith, Saul de Leon—whom she encounters when she attends her church for the first time in almost fifteen years—emerges as a more unexpected mentor. The son of a rabbi and one of Arnie's classmates in high school, Saul is a Christian convert. In response to Mum Hattie's blunt questions, Saul traces in intricate detail the steps that took him from his view of Jesus as a "Jewish social revolutionary" to the miracle of his conversion to Christianity. Saul emerges, in effect, as Anne's pastoral counselor, guiding her through their frequent conversations to a deeper understanding of her political and spiritual failings:

> I understood more clearly how a nation could become Godless as I reviewed my own experience of the past years when I had, one by one, abandoned practices and habits of though and of belief, and had accepted others in their place. Now I found that each day I must travel that road again, back the way I had come, submitting to the Creator all the ideas that had possessed my mind for so long. It was not easy to do this. It was painful to relinquish each treasured Socialist belief and set another in its place; to love God with my mind as well as with my heart and soul.[53]

To the ménage consisting of Anne, Arnie, Mum Hattie, and Saul de Leon is added Anne's former husband, David, now a disheveled, financially insolvent alcoholic whom Anne reluctantly allows to live in her house. He, too, will play a role in the denouement of the novel.

Saul de Leon brings to Anne's attention a service that had been used by the Church in the fourth century:

> It was a service for exorcising demons, and the possessed persons were called "energumens." After an exhortation by the Bishop, beginning, "Pray, ye energumens who are vexed with unclean spirits..." and certain prayers, the Bishop, in the Name of the Trinity, boldly ordered the demons to leave the possessed bodies.[54]

Saul convinces Anne that David still possesses the priestly authority to ad-minister this ceremony, and, rousing David from his daily alcoholic stupor, Anne, Mum Hattie, and Saul join with him in the ritual, an exorcism of sorts, that restores Arnie—and Anne—to their senses.

Anne Braxton has indeed traveled "full circle," recovering in the process the values, beliefs, and habits of thought that she now claims as her birth-right. In her step-by-step repudiation of the political values that have shaped her life for the past twelve years or so, however, she has also rejected the so-cial concerns that brought her into the orbit of the Communist Party in the first place, particularly the struggle for racial justice embodied in the case of two young African Americans falsely accused of rape. For Anne Braxton—and, apparently, for Grace Lumpkin—the perfidy of the Communists auto-matically discredits any of the causes with which they are, or have been, asso-ciated, since all of their causes are ultimately subordinate to the larger goal of undermining and destroying the religious basis of American life. By this con-voluted logic, *anything* associated with the Communist Party must be deci-sively rejected.

The Scottsboro case had been seized upon by the Communist Party, to be sure, but it had not been created by it. Nevertheless, as the details of the case began to recede from public consciousness (and there had been at least two "official" conclusions to the case by the early 1960s: the release of four of the Scottsboro defendants—Roy Wright, Olen Montgomery, Willie Roberson, and Eugene Williams—in 1937; and the parole of the last Scottsboro still in prison, Andy Wright, in 1950), the casual conflation of Scottsboro and Com-munism continued to circulate in American culture.

Although neither Don Mankiewicz's *Trial*, as novel or film, nor Grace Lumpkin's *Full Circle* can be said to have significantly molded public opin-ion during the 1950s, they are symptomatic of the forces that helped to con-tribute to the cultural amnesia surrounding the Scottsboro spectacle.

CHAPTER EIGHT

HARPER LEE'S *TO KILL A MOCKINGBIRD*

The Final Stage of the Scottsboro Narrative

"In our courts when it's a white man's word against a black man's, the
white man always wins. They're ugly, but those are the facts of life."

—Atticus Finch[1]

FOR ALL OF THE VARIOUS forms the Scottsboro Narrative took as it traveled
through American culture, it achieved its most stubbornly enduring incarnation in Harper Lee's now classic novel *To Kill a Mockingbird*, and, even
more potently, in the film based upon the novel. To call these inextricably
connected works the final stage of the Scottsboro Narrative is not to deny
that the historical case continues to engage the imagination of artists, playwrights, novelists, and filmmakers;[2] it is simply to acknowledge that in early
twenty-first-century America, *To Kill a Mockingbird* is more and more frequently taught, if not read, in creative tension with the 1931 Scottsboro
case[3]—and that, for many Americans, their encounter with the case occurs
through the pages of Lee's novel, or through the images of the film—even
when they are not aware of it.

This was not initially the case, and it is striking how quickly the novel,
once condescendingly described as "pleasant, undemanding reading" by
Phoebe Adams in the *Atlantic* magazine when it was first published in 1960,[4]
has come to occupy such an extraordinarily influential position in American
and global culture. In its first year of publication it shattered many records
in publishing history: selected by three American book clubs (the Book-of-
the-Month Club, the Literary Guild, and Reader's Digest Condensed Books),
it sold more than two and a half million copies and went through fourteen
printings. It was also chosen by the British Book Society, and was published
in France, Germany, Italy, Spain, Holland, Denmark, Norway, Sweden,
Finland, and Czechoslovakia.[5] Awarded the Pulitzer Prize in Letters in May
1961, *To Kill a Mockingbird* had sold more than five million copies in 1962,
eleven million copies in 1975, over fifteen million by 1982, and over eighteen million paperbacks by 1992.[6] According to a 1991 "Survey of Lifetime
Reading Habits" conducted by the Book-of-the-Month Club and the Library

of Congress's Center for the Book, *Mockingbird* was regarded by its 5,000 respondents as one of three most influential books in people's lives, second only to the Bible.[7] After the novel was made into a film in 1962, winning three Oscars, including Best Actor for Gregory Peck, the novel and the film became inextricably linked in the public imagination and *Mockingbird's* cultural influence became even more pervasive—permanently enshrining Gregory Peck as an icon of dignified, upstanding, ethical behavior and establishing the fictional southern lawyer he portrayed, Atticus Finch, as the number one hero in American movie history, according to a recent American Film Institute survey.[8] How to account for the extraordinary longevity of *To Kill a Mockingbird* and its enduring popular appeal?

There is intrinsic pleasure, to be sure, in the retrospective account of a young white woman growing up in a sleepy southern town populated by eccentrics in the midst of Depression-era America. But this somewhat idyllic portrait is bracketed by the childhood terrors provoked by the shadowy presence of Boo Radley and the palpably unjust trial of Tom Robinson, an African American accused of rape by a lower-class white woman. In short, *To Kill a Mockingbird* is deeply concerned with issues of race. Surely Eric Sundquist is right when he observes that the continuing popularity of *To Kill a Mockingbird* presents an intriguing paradox: "the novel is the most widely read twentieth-century American work of fiction devoted to the issue of race," yet it has become "an icon whose emotive sway remains powerful because it also remains unexamined."[9] Harper Lee's treatment of racial issues is subtle, and is often displaced and deflected by a narrative point of view that skillfully blends the perspective of a mature adult reflecting back on her experiences with that of a child who recounts those experiences as they occur. In spite of the apparently simple and direct story that Scout tells, through her Lee offers a major revisionist perspective on Alabama history—past and present—at the heart of which she offers her own interpretive framework for the Scottsboro case.

Eric Sundquist has persuasively argued that *To Kill a Mockingbird* can usefully be read as an allegory of the historical moment of its publication, casting itself imaginatively backward in time in order to reflect, from a safe distance, upon the anxieties and racial tensions engendered by the 1954 Supreme Court decision *Brown v. Board of Education* and the growing momentum of the modern Civil Rights Movement: "The novel harks back to the 1930s both to move the mounting fear and violence surrounding desegregation into an arena of safer contemplation and to remind us, through a merciless string of moral lessons, that the children of Atticus Finch are the only hope for a future world of racial justice."[10] But I would argue that *To Kill a Mockingbird* is also deeply preoccupied with how the past is viewed and that the Scottsboro case, for Lee, remains an important touchstone for gauging the potential of the South for civilized behavior.

From the beginning, discerning critics noted the parallels between Harper Lee, born Nelle Harper Lee, the youngest of four children, on April 28, 1926 in Monroeville, a small town in southwest Alabama, and her fictional creation, Scout Finch. Her father, Amasa Coleman Lee, was a lawyer who served in the Alabama state legislature 1926–38; her mother, Frances Cunningham Finch Lee, was a member of a distinguished Virginia family that founded Finchburg, Alabama, in Monroe County. A precocious child and a tomboy, Lee counted among her closest childhood companions her older brother, Edwin Coleman Lee, and her cousin, Truman Capote, who was her next-door neighbor from 1928 to 1933. Harper Lee was the inspiration for Capote's character Idabel in his *Other Voices, Other Rooms* (1948), just as he was the prototype for the character Dill in *To Kill a Mockingbird*. She attended public school in Monroeville, attended Huntington College, a private school for women in Montgomery, from 1944 to 1945, then transferred to the University of Alabama. There she studied law between 1945 and 1949, of which she spent a year as an exchange student at Oxford University. In 1949 she abandoned her studies, moving to New York City, where she worked as a reservations clerk for Eastern Air Lines and British Overseas Airways while she pursued her literary ambitions. She wrote a series of short stories about Monroeville, received encouragement for them from a literary agent, and— with the generous financial support of a group of close friends—quit her job so that she could devote her full attention to her writing. In 1957 she submitted a first draft of *To Kill a Mockingbird* to Tay Hohoff, an editor at J. B. Lippincott, who saw enough promise in it to support Lee through three years of revisions—interrupted by periodic visits back to Monroeville to visit her ailing father.[11] *To Kill a Mockingbird* was published in 1960—and Harper Lee triumphantly marched into publishing history.

Although many critics have noted the parallels between Harper Lee and Scout Finch, it has only been within the past decade or so that they have begun to emphasize the striking parallels between the trial of Tom Robinson—the heart of the novel—and the historic Scottsboro trials. The precocious Harper Lee was five years old when the first Scottsboro trials occurred, eleven when the first four defendants were finally released, leading some critics to speculate that Lee was deeply, and permanently, affected by these events.

To simply note the striking parallels between Tom Robinson's trial and those of the Scottsboro boys, however, is to beg the question of Harper Lee's conflation of these events in her novel, and her particular rendition of a "Scottsboro Narrative."

To Kill a Mockingbird occurs within a carefully delineated historical framework, between the years 1932 and 1935 (roughly the peak years of the Scottsboro trials) in the Depression-era small town of Maycomb in southwest Alabama. There is widespread unemployment throughout the country and Maycomb is feeling its effects, with professionals like Atticus Finch being paid for their services in promises and produce rather than cash. There are occasional hints of social upheaval in the cities of Alabama, but they seem removed from the world occupied by Scout and her brother Jem: "there were sit-down strikes in Birmingham; bread lines in the cities grew longer, people in the country grew poorer. But these were events remote from the world of Jem and me."[12]

Strikingly, after his encounter with a threatening mob outside of the jail where Tom Robinson is held, Atticus Finch is adamant in his insistence to Scout and Jem that the Ku Klux Klan does not exist in Alabama:

"... we don't have mobs and that nonsense in Maycomb. I've never heard of a gang in Maycomb."

"Ku Klux got after some Catholics one time."

"Never heard of any Catholics in Maycomb, either," said Atticus, "you're confusing that with something else. Way back about nineteen-twenty there was a Klan, but it was a political organization more than anything. Besides, they couldn't find anybody to scare. They paraded by Mr. Sam Levy's house one night, but Sam just stood on his porch and told 'em things had come to a pretty pass, he'd sold 'em the very sheets on their backs. Sam made 'em so ashamed of themselves they went away.". . .

"The Ku Klux's gone," said Atticus. "It'll never come back."[13]

Excising the Ku Klux Klan from her fictional landscape is only one of the ways in which Lee constructs the anatomy of Maycomb. Although threats to world peace loom—there are occasional allusions to the rise of Adolf Hitler in Germany and the persecution of Jews there—the universe of Maycomb, Alabama is depicted as more or less intact. To be sure, there are sharply defined social distinctions and racial divisions that are carefully observed by its inhabitants. If trouble comes to Maycomb, the narrative makes plain from the very beginning, it will be caused by "outsiders," those people who exist outside of the intricate etiquette that binds Maycomb's residents together. Moreover, "Atticus was related by blood or marriage to nearly every family in the town."[14] It is from this relatively privileged position that Scout narrates the experiences that will catapult her, Jem, and Dill into maturity.

Wide-eyed and precocious, Scout is a tomboy locked into an ongoing battle with the behavior expected of a young white woman of her social background; she is in constant rebellion against "the starched walls of a pink cotton penitentiary closing in on me."[15] Although Scout's rebelliousness is sometimes curbed by the stern but loving hand of the Finch's African American housekeeper Calpurnia, sometimes by her aunt Alexandria, Atticus's sister, it is often abetted by Atticus, her widowed father, an apparently endless source of wisdom about how to conduct oneself in a community like Maycomb. Scout's embattled consciousness thus offers a particular angle of vision towards the society she inhabits, providing along the way a clinical view of the forces that shape it.

The first and most important lesson that Scout must learn is that of tolerance. As Atticus stresses in one of his homilies to her at the beginning of the novel: "You never really understand a person until you consider things from his point of view . . . until you climb in his skin and walk around in it."[16] Atticus's insistence upon the fundamental principle of fair play is underscored by periodic references in the novel to the real world phenomena of "Seckatary Hawkins," the fictional persona of Robert Franc Schulkers, founder of the Fair & Square Club—to which Harper Lee belonged as a child. Founded in 1920 and centered around books about a clever little fat boy and his friends, the Fair & Square Club sought to inculcate among its members "a 'can-do' spirit, harmonized with principles of God & family, friendship, fair-play, equality, and patriotism." Its detailed list of "Club Rules" concluded with the following membership pledge: "I shall always be fair and square, possessed with strength of character, honest with God and my friends, and in later life, a good citizen."[17] In an important sense, a great deal of the action of *To Kill a Mockingbird* revolves around the testing of these principles of tolerance and fair play against the characters farthest removed from the world inhabited by Scout, Jem, and Dill: Boo Radley, Tom Robinson, and, arguably, the Ewell family.

When the action of *To Kill a Mockingbird* shifts from the world of childhood terror, symbolized by Scout, Jem, and Dill's morbid fascination with their reclusive neighbor, Boo Radley, to the trial of Tom Robinson, the mood of the novel noticeably shifts as well: the children are now forced to confront a bewildering world of legal and social codes fueled by racial undercurrents they only dimly understand.

In her construction of Tom Robinson's case, Harper Lee clearly could have drawn upon a wide range of historical antecedents, in Alabama and other parts of the South, including the 1930 case of a black Alabaman named Tom Robinson who was convicted and sentenced to death for defending his family from a lynch mob,[18] but her imagination was in the grips of the Scottsboro case.[19]

The clearest parallels between the Scottsboro case and Tom Robinson's trial revolve around the circumstances themselves: the false accusation of rape by a lower-class southern white woman. Mayella Ewell's accusation provokes a chain reaction of events that causes Atticus to fret: "I hope and pray I can get Jem and Scout through it without bitterness, and most of all, without catching Maycomb's usual disease. Why reasonable people go stark raving mad when anything involving a Negro comes up, is something I don't pretend to understand."[20] The "usual disease" initially leads to a mob gathering reminiscent of the one that greeted the Scottsboro defendants at Paint Rock, Alabama—even though, as readers of the novel know, Scout skillfully and single-handedly defuses the mob's threat. The rule of law prevails, however, as it did during the Scottsboro trials—regardless of how flawed those rules may have been. Atticus defends Tom Robinson because he is appointed by the court to do so, but he is also compelled by his ethical commitments. As he tells Scout: "If I didn't I couldn't hold up my head in town, I couldn't represent this county in the legislature, I couldn't even tell you or Jem not to do something again."[21] Atticus is considerably more skilled, and respected, than the somewhat disreputable Stephen Roddy, the attorney first appointed to defend the Scottsboro Boys, but he is acutely aware of the social and legal forces that have been mobilized against him. In fact, he informs Scout long before the trial begins that he will not win the case: "Simply because we were licked a hundred years before we started is no reason for us not to try to win."[22] Besides, Atticus continues: "This time we aren't fighting the Yankees, we're fighting our friends. But remember this, no matter how bitter things get, they're still our friends and this is still our home."[23]

This is, in a sense, a Faulknerian conundrum. Atticus's rationale in this passage dramatizes Harper Lee's sharpest departure from imaginative renderings of the Scottsboro case. There is no Helen Marcy here, no Hollace Ransdall; no organizers from the International Labor Defense appear in this novel, no ILD lawyers, no Walter White and William Pickens, no NAACP, no reporters from the national press, no national and international protests, no U.S. Supreme Court hovering in the background. The case of Tom Robinson is a *local* matter, Atticus insists, a family feud—and Harper Lee apparently assents to this perspective in her handling of her fictional material. Atticus is serenely confident in his belief that the citizens of Maycomb are basically good. In this regard he seems to share some affinities with George Chamlee, the Tennessee lawyer from a distinguished southern family who was hired by the ILD to defend the Scottsboro Boys partly because of his deep ties to the region, as well as with Judge James Horton, the courageous judge who set aside Haywood Patterson's second conviction in 1933 and ordered a new trial—destroying his political future in the process.[24]

Tom Robinson's trial is preceded by tensions between Scout, Jem, and some of their classmates at school as well as with some of the residents of Maycomb, conflicts which, in turn, spill into the Finch household, as Scout, Jem, and Atticus, in effect, debate social assumptions and legal tactics—sometimes heated discussions that, in fact, reveal sharp differences among them. After Scout disarms the mob that has gathered at the Maycomb jail to exact vigilante justice upon Tom Robinson by drawing into conversation one of its leaders, Walter Cunningham, she pointedly remarks to Atticus that Cunningham would have hurt him, given the opportunity; Jem supports her argument:

Atticus placed his fork besides his knife and pushed his plate aside. "Mr. Cunningham's basically a good man," he said, "he just has his blind spots along with the rest of us."

Jem spoke. "Don't call that a blind spot. He'da killed you last night when he first went there."

"He might have hurt me a little," Atticus conceded, "but son, you'll understand folks a little better when you're older. A mob's always made up of people, no matter what. Mr. Cunningham was part of a mob last night, but he was still a man. Every mob in every little Southern town is always made up of people you know—doesn't say much for them, does it?"

"I'll say not," said Jem.

"So it took an eight-year-old child to bring 'em to their senses, didn't it?" said Atticus. "That proves something—that a gang of wild animals *can* be stopped, simply because they're still human. Hmp, maybe we need a police force of children . . . you children last night made Walter Cunningham stand in my shoes for a minute. That was enough."[25]

Never one to miss an opportunity for moral instruction, Atticus insists upon maintaining his essentially optimistic view of human nature and his fellow citizens. As the final chapters of the novel suggest, however, Jem's point of view—and, perhaps, Scout's as well—may be more consistent with the available evidence.

Stripped down to its bare essentials, the trial of Tom Robinson has none of the tensions or fireworks of its historical prototype; rather, it takes on the leisurely, sleepy dimensions of the town of Maycomb itself, complete with running commentary from Scout and Jem, comfortably settled with Maycomb's black residents in the balcony of the courtroom. From the beginning of the trial Scout's commentary is marked by her palpable disdain for the Ewell family, dramatizing the social antagonisms that have been latent since the beginning of the novel:

> Every town the size of Maycomb had families like the Ewells. No economic fluctuations changed their status—people like the Ewells lived as guests of the county in prosperity as well as in the depths of a depression. No truant officers could keep their numerous offspring in school; no public health officer could free them from congenital defects, various worms, and the diseases indigenous to filthy surroundings.[26]

The Ewells live "behind the town garbage dump in what was once a Negro cabin." There can be no lower social status for whites in the terms of Maycomb's universe. Robert E. Lee Ewell, the father of the complainant, is "a little bantam cock of a man" and a literal red-neck to boot; his daughter, Mayella Violet, initially appears vulnerable, but Scout's subsequent description of her quickly withdraws the possibility of sympathy: "she seemed somehow fragile-looking, but when she sat facing us in the witness chair she became what she was, a thick-bodied girl accustomed to strenuous labor."[27]

Thus, through Scout's observations, does Harper Lee confront one of the strategic issues that faced the original defenders of the Scottsboro Boys: how to characterize the motives and behavior of their accusers, who shared the same general social and economic plight? Some of the early Scottsboro defenders had found a convenient bifurcated solution to the question after February 1932, when Ruby Bates recanted her testimony and was taken up by the Communist Party: Ruby Bates became reconstituted as a "worker" who joined some of the Scottsboro mothers on speaking tours, while Victoria Price adamantly clung to her accusations, regaling one jury after another with the sordid tale of her supposed assault and justifying the view of her among many of the Scottsboro supporters as an unscrupulous, lying prostitute. In Scout's hands, Mayella Ewell, too, falls outside the pale of human sympathy. Although she is clearly terrified by her courtroom encounter with Atticus, and wary of him, she is also sullen, resentful, sometimes inarticulate, and wrathful. Moreover—and most importantly—Atticus's careful cross-examination exposes her as a sexual aggressor who may have been abused by her father. These revelations, however, do not mitigate the moral judgments against her, judgments issued explicitly by Atticus in his summation to the jury:

> "I have nothing but pity in my heart for the chief witness for the state, but my pity does not extend so far as to her putting a man's life at stake. . . .
>
> "She has committed no crime, she has merely broken a rigid and time-honored code of our society, a code so severe that whoever breaks it is hounded from our midst as unfit to live with. She is the victim of cruel poverty and ignorance, but I cannot pity her: she is white. She knew full well the enormity of her offense, but because her desires were stronger than the code she was breaking, she persisted in breaking it. . . . she struck out at her victim—of necessity she must put him away from her—he must be removed from her presence, from this world. She must destroy the evidence of her offense."[28]

In sharp contrast to his accusers—and to his historical antecedents—Tom Robinson is a model of restraint and decorum. Twenty-five years old, married, and the father of three children, Tom Robinson is older than the oldest Scottsboro defendant—and more reputable. In physical terms, he is a "black, velvet Negro. . . . If he had been whole, he would have been a fine specimen of a man."[29] But Tom is afflicted by a visible physical impairment: "His left arm was fully twelve inches shorter than his right, and hung dead at his side. It ended in a small shriveled hand, and from as far away as the balcony I could see that it was no use to him."[30] Tom Robinson's physical deformity makes him no more likely a suspect in the supposed rape of Mayella Ewell than did Olen Montgomery's half blindness and Willie Roberson's advanced venereal disease in the story manufactured by Victoria Price in her testimony during the Scottsboro trials.

Interestingly enough, Atticus's exposure of Mayella Ewell as the intiator of physical contact with Tom Robinson, and Robinson's description of his aggrieved innocence in the face of such behavior, enters territory that even the Scottsboro Boys' lawyer during the Decatur trials, Samuel Leibowitz, would not imagine considering. Leibowitz cast doubt on Victoria Price's morals, to be sure, but there was no hint in his aggressive examination of her that she had willingly bestowed her sexual favors upon any of the Scottsboro Boys. Through his skillful and courtly examination of Mayella, Atticus establishes not only that she craved a sexual encounter with Tom Robinson, but that she plotted for over a year to achieve her desires. In short, Atticus skillfully dissects Mayella and her way of life—and Mayella, and everyone else in the courtroom, knows it. The only outcome of Atticus's brilliant defense, however, is that he manages to delay the inevitable outcome of the trial for several hours, as the jury assembles to consider its verdict—a pyrrhic victory. As Miss Maudie comments the day after the trial:

> ". . . as I waited I thought, Atticus Finch won't win, he can't win, but he's the only man in these parts who can keep a jury out so long in a case like that. And I thought to myself, well, we're making a step—it's just a baby-step, but it's a step."[31]

After the trial another argument rages in the Finch household. While Tom Robinson is held in prison seventy miles away, and his wife and children are denied permission to visit him, Atticus assures his children that Tom has a good chance of winning on appeal. Atticus supports the death penalty in rape cases, it turns out, but believes the judgment should be based on solid, not circumstantial, evidence. Jem is not so sure, and argues for his own version of jury nullification: "'Then it all goes back to the jury, then. We oughta do away with juries.' Jem was adamant."[32] Had this argument prevailed during the 1933 Scottsboro trials in Decatur, the Scottsboro Boys probably would have been set free. Atticus counters, however, that such a change would require legislative approval, and changing the law is a slow and onerous process: "You'd be surprised how hard that'd be. I won't live to see the law changed, and if you live to see it you'll be an old man."[33] Atticus's "wisdom" in this instance, his rejoinder to Jem's desire for radical change in the legal system, is tantamount to support for the status quo, rooted in the view of human nature he has unswervingly upheld throughout the novel: "Those are twelve reasonable men in everyday life, Tom's jury, but you saw something come between them and reason."[34] That "something" is the same as "Maycomb's usual disease"—racism—and Atticus's failure, or unwillingness, to recognize the tragic consequences of this "flaw" among his fellow townspeople exposes the limitations of his imagination and wisdom.

The assumption of the novel is that Tom Robinson is already receiving the best legal representation possible in Atticus Finch, so the issue of proper legal counsel that led to the reversal of the first Scottsboro convictions in the historic 1932 Supreme Court decision *Powell v. Alabama* is rendered moot. But Jem, still not satisfied, turns his attention to the composition of the jury:

"Why don't people like us and Miss Maudie ever sit on juries? You never see anybody from Maycomb on a jury—they all come from out in the woods."[35] Women can't serve on juries, Atticus patiently explains, much to Scout's indignation—and thereby implicitly negates the basis for the second major Supreme Court ruling on the Scottsboro case—the routine practice of excluding African Americans from juries in Alabama, as cited in the 1935 ruling *Norris v. Alabama* that reversed the second conviction of Clarence Norris and the third of Haywood Patterson.

The question of Tom Robinson's right to be tried by a jury of his peers is, apparently, outside of the realm of Atticus's imagination, too. He then shifts the grounds of the discussion by casually revealing that one of the twelve jurors initially voted for an outright acquittal of Tom Robinson—one of the extended Cunningham clan. He thus reintroduces the issue of social class, but in such a way as to throw both Scout's and Jem's assumptions into confusion. As Scout and Jem prepare for bed, Jem tries to authoritatively sum up what they have learned—from the trial, and from Atticus: "I've thought about it a lot lately and I've got it figured out. There's four kinds of folks in the world. There's the ordinary kind like us and the neighbors, there's the kind like the Cunninghams out in the woods, the kind like the Ewells down at the dump, and the Negroes."[36] The Chinese and "the Cajuns down yonder in Baldwin County"—all outsiders—don't count, he solemnly informs Scout, just the people who inhabit Maycomb County. Scout remains skeptical: "Naw, Jem, I think there's just one kind of folks. Folks."[37] Jem's rejoinder hovers over the rest of the novel: "'That's what I thought, too,' he said at last, 'when I was your age. If there's just one kind of folks, why can't they get along with each other? If they're all alike, why do they go out of their way to despise each other?'"[38]

Conversations such as these contribute to a general sense of helplessness, summed up most poignantly by Dill, who announces after the trial: "I think I'll be a clown when I get grown. . . . Yes sir, a clown. . . . There ain't one thing I can do about folks except laugh, so I'm gonna join the circus and laugh my head off."[39]

In the immediate aftermath of the trial, Scout must undergo yet another ordeal (and, needless to say, the entire Tom Robinson episode is based upon Scout's experience and *crise de conscience*, not Tom Robinson's—thus firmly locating *To Kill a Mockingbird* in a literary and popular tradition at least as old as Mark Twain's *Huckleberry Finn*, wherein the deep-rooted issues of race and racism achieve their fullest expression in the psyches of young white protagonists): the meeting of her Aunt Alexandria's missionary circle, where many of its members spout empty platitudes about the remote Mruna people while cruelly berating their African American domestics for their sullenness about the Robinson trial. Once again, Scout sees through the entrenched hypocrisy of her society, and she quickly distances herself from it: "Ladies in bunches always filled me with vague apprehension and a firm desire to be elsewhere."[40] She also privately reaffirms her commitment to justice for Tom Robinson: "I wished I was the Governor of Alabama for one day; I'd let Tom Robinson go so quick the Missionary Society wouldn't have time to catch its breath"[41]—dreams that are of course dashed by Atticus's unexpected appearance at home to inform his family that Tom Robinson has been killed in a futile attempt to escape from prison.

After Atticus leaves with Calpurnia to inform the Robinson family of his death, Scout, Aunt Alexandria, and Miss Maudie take momentary refuge in the sanctuary of the Finches' kitchen, away from the prying eyes and ears of the missionary circle. Miss Maudie—who throughout the novel has functioned as a surrogate mentor for Scout—explodes, paying tribute to Atticus's heroism at the same time that she castigates the moral cowardice of those townspeople who share his values and beliefs: "'The handful of people who say a fair trial is not marked White Only; the handful of people who say a fair trial is for everybody; the handful of people with enough humility to think, when they look at a Negro, there but for the Lord's kindness am I. . . . The handful of people in this town with background. . . .'"[42] If these people, the people with background, had stood up for their convictions and assumed their proper responsibilities as citizens, presumably the outcome of the trial would have been different.

Tom Robinson's death and Atticus's meeting with his family both occur offstage, so to speak, outside of the immediacy of Scout's experiences, so her responses to these events are really shaped by secondhand accounts. As she sifts through the various reports, including the impassioned editorial that appears in the *Maycomb Tribune*—which compares Tom Robinson's death to "the senseless slaughter of songbirds by hunters and children"—she offers the following final assessment of the case: "Senseless killing—Tom had been given due process of law to the day of his death; he had been tried and convicted by twelve good men and true; my father had fought for him all the way. . . . Atticus had used every tool available to free Tom Robinson, but in the secret courts of men's hearts Atticus had no case. Tom was a dead man the minute Mayella Ewell opened her mouth and screamed."[43] This is virtually her last word on the subject before the end of the novel and her final, fateful encounters with Bob Ewell and Boo Radley.

In the final analysis, the parallels between the Scottsboro case and Tom Robinson's trial rest not only in the details, which Harper Lee clearly reshapes for her own fictional purposes, but in her clinical inventory of the social and historical forces that shaped Alabama responses to the Scottsboro case—attitudes and behavior that she replicates in her characters. By avoiding any references to the intense national and international scrutiny that was trained on the Alabama during the 1930s, Lee not only turns the case into a strictly local matter, she strongly suggests, in fact, that it was best understood or treated as such. In the folklore that still circulates in Scottsboro, Alabama, some of the residents who remember the case sometimes still say about the Scottsboro Boys: "It wasn't our niggers and it wasn't our whores,"[44] meaning that local African Americans familiar with the prevailing racial etiquette would never have behaved the way the Scottsboro Boys presumably did—nor would local white prostitutes, for that matter; *and*, more ambiguously, that Scottsboro residents would have treated *their* blacks differently than they did the Scottsboro Boys. "Outside interference," the long-standing mantra of many southern defenders of the racial status quo, would only muddy the waters. Cases such as Tom Robinson's—and the Scottsboro case—were best handled by people armed with an intimate understanding of the mores of the community, such as Atticus Finch. This kind of specious reasoning, Byron Woodfin noted in another context, "would be the same argument later used to justify the bloody and deadly violence when such groups as the Freedom Riders converged on the South during the decisive civil rights battle."[45] Through Scout's careful observations, Lee indicts those values that perpetuate prejudice and racial intolerance at the same time that she fiercely upholds the traits of gentleness, humility, dignity, and empathy embodied in the figure of Atticus Finch—even if *To Kill a Mockingbird* unwittingly exposes the weakness of those values in the face of widespread and deeply entrenched social and political injustice.

Screenwriter Horton Foote's adaptation of Lee's novel and Robert Mulligan's direction carefully chisel away the multiple dimensions of Scout's first-person narrative and the complex terms of her universe, focusing the film more directly upon Atticus Finch and placing the trial of Tom Robinson squarely at its dramatic center.[46] Whereas Robinson's trial occupied about fifteen percent of the novel, it consumes about one-third of the film, underscoring its deep immersion in the racial politics of the early 1960s.[47] Equally significant was the decision to cast Brock Peters as the ill-fated Tom Robinson.

Nearly forty years after the film was released, at a gathering in Wichita, Kansas with Mary Badham and Phillip Alford called "Scout Remembers Mockingbird," Brock Peters could still sharply recall his feelings about winning the role:

I found myself the last of two people being considered for this role and of course it was a very difficult time, because I thought, *I've come this far, surely they're going to let me have this role.* And I was really worried because my competition was one of our finest actors: James Earl Jones. That's the pressure we often have to live under if we know who our competition is. . . . My agent called me and said *we have a meeting for you* to talk with the producers, the director and all those concerned, and after this meeting a decision will be made. Well of course I was scared out of my wits. I didn't know how to present myself in order to get this coveted prize. I went into the meeting—it was in a building at Park Avenue and 57th Street and I

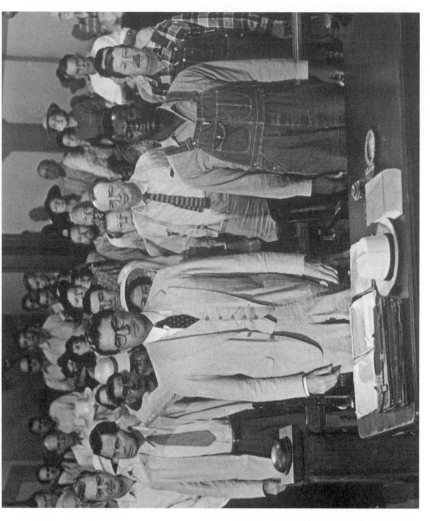

Figure 8.1 Scene from *To Kill a Mockingbird*. Source: MPTV Images.

tried not to appear frightened but I wanted to look cool and calm and still suggest the character of Tom Robinson, and do that dressed in a suit.

I came away from that meeting and within a matter of days they gave me the nod. And I was really very happy about that. I'm not so sure that Jimmy was. Subsequently we have come head-to-head on a couple of other projects and he has won and I keep saying *let's stop this. We don't need to do this Jimmy.*[48]

In some respects, Peters's triumph was a pyrrhic victory. As Donald Bogle, the irrepressible historian of black images in American cinema, has noted, Peters was "the era's most prominent black brute figure."[49] In a milieu populated by such actors as James Earl Jones, Woody Strode, Harry Belafonte, Sidney Poitier, Ivan Dixon, James Edwards, Ossie Davis, and Bernie Hamilton, Peters, Bogle notes, was the personification of the classic black brute. He had snarled his way through the role of Sargeant Brown in his 1954 film debut in *Carmen Jones* and he had played the malevolent Crown in the 1959 film version of *Porgy and Bess*: "Brock Peters was pitch black (that was enough to frighten any white audience) and without the healthy good looks of a Poiter or a Belafonte. He perfected a vicious snarl and as he malevolently eyed the other characters in the film he suggested uncontrolled violence. Finally, when he pounced upon his opponents, he was the incarnation of unbridled brutality."[50]

Although Peters's screen persona would soften somewhat when he was cast as the West Indian homosexual Johnny in *The L-Shaped Room* the following year, he was still pretty unreconstructed in 1962. It was almost as if *To Kill a Mockingbird* could not resist flirting with the very racial stereotypes, particularly the threat of sexual menace represented by black men, it presumably sought to subvert. With his flaring nostrils and smouldering rage, Peters dramatically conveyed Tom Robinson's sense of aggrievement at the same time that the film suggested that it was Atticus's moderation and humanity that held his deepest emotions in check. Try as he might, Peters could not quite conceal the virility of his stage presence—his physical affliction notwithstanding. Peters's portrayal of Tom Robinson is, in short, the perfect foil for Atticus Finch and it underscores the extent to which, in the final analysis, his destiny is nowhere as significant as that of Atticus, Scout, and Jem.

Like the novel upon which it is based, the film version of *To Kill a Mockingbird* suggests that the solution to apparently untractable social problems ultimately rests in the hands of enlightened individuals, like Atticus Finch; indeed, the problems exists as a means of developing and refining the moral awareness of characters like Scout and Jem who will turn their attention to them in due time. Although Atticus is soundly defeated by the inequities of the legal system and the forces of racism in his community, he nonetheless maintains his moral integrity and is allowed to cling to his deep belief that most people are basically decent when viewed on their own terms. Moreover, within the hierarchy of childhood terrors both the novel and the film create, Tom Robinson's trial and subsequent death is displaced by the very real threat presented by Bob Ewell at the end of the novel.

Bob Ewell's death at the hands of Boo Radley signifies the expulsion of evil from Maycomb and also represents an unintentional act of atonement for Tom Robinson's death, a connection Sheriff Tate explicitly makes to Atticus:

> "Bob Ewell fell on his knife. He killed himself. There is a black man dead for no reason, and now the man responsible for it is dead. Let the dead bury the dead this time, Mr. Finch. . . . maybe you'll tell me it's my duty to tell the town all about it, not to hush it up. . . . To my way of thinkin', takin' one man who's done you and this town a big service, and draggin' him, with his shy ways, into the limelight, to me, that's a sin. It's a sin, and I'm not about to have it on my head."[51]

After reflecting upon the Sheriff's argument, Atticus enlists Scout into the cover-up, and Scout, startling him, quickly accedes because, as she tells him, to implicate Boo Radley in Bob Ewell's death would "be sort of like shootin' a mockingbird, wouldn't it?"[52]

In spite of Atticus Finch's often stated faith in the judicial system, it is worth noting that the rough justice meted out by Boo Radley falls squarely outside the realm of established authority. As tidy and symmetrical as the ending of *To Kill a Mockingbird* may be, it dramatizes how justice, in the final analysis, becomes a private matter—and one that rests firmly upon tacit agreements among white men. The full magnitude of Tom Robinson's trial and death must necessarily remain offstage. This ending settles individual scores, but it leaves existing social hierarchies and systemic injustices intact.

To Kill a Mockingbird represents the last stage of the Scottsboro Narrative in the sense that it seems to mark the outer limits of what many readers and viewers are prepared to accept in the ongoing dialogue and debate about race and justice in the United States. In spite of the passions unleashed by the historic Scottsboro case, the angry protests of thousands upon thousands of people, in the United States and abroad, and the powerful imagery of the Scottsboro Narrative that was forged during the heyday of the case, that Scottsboro has receded from American historical memory. The version of Scottsboro that has become deeply entrenched in the American popular imagination now rests securely in *To Kill a Mockingbird*.

EPILOGUE

IF THE SCOTTSBORO NARRATIVE constituted such a significant dimension of American racial discourse between 1931 and the 1970s, if it has emerged the touchstone by which subsequent miscarriages of justice can be measured, what are we to make of its relative disappearance from public discourse in more recent times; or—to pose the question in another way—what are we to make of its periodic *reappearance* in American racial discourse?

In recent years, the Scottsboro case has been invoked in the context of three traumatic national racial spectacles: the 1989 Central Park jogger rape case, the 1994–95 O. J. Simpson trial, and the 2007 Duke University rape case.

Ironically, the case that has the closest parallels to Scottsboro is that of the Central Park jogger, an event so shocking in its brutality and so harrowing in its details that it initially defied any comparisons. On April 19, 1989, a twenty-eight-year-old white woman, a Phi Beta Kappa graduate of Wellesley and Yale, an investment banker at Salomon Brothers who was rapidly rising through the ranks, went for an evening run in New York City's Central Park. Shortly thereafter she was stripped of her clothes, raped, and viciously assaulted—so badly that one of her eye sockets was crushed and her skull smashed. By the time the police arrived she had lost 80 percent of her blood, and the doctors who attended her at Metropolitan Hospital were very pessimistic about her possibilities of survival. After regaining consciousness the victim had absolutely no memory of the attack.[1]

The brutal attack on the Central Park jogger occurred on a night of random violence during which a roving gang of about forty African American and Latino teenagers entered Central Park shortly after nine p.m. with the express purpose, according to some of them, of assaulting and robbing people. Not everyone in the loose group knew everyone else, and the group split and re-formed as it terrorized people during the next hour. Other joggers, cyclists, pedestrians, and motorists were threatened or assaulted.[2] These random attacks introduced the term "wilding" into the national lexicon as a signifier of urban nihilism, particularly in a New York City which, in the late 1980s, was racked with social, economic, and racial tensions.

Shortly after the widespread attacks in the park were reported, the police swept through and rounded up many of the rampaging teenagers, including two of the subsequent defendants in the Central Park jogger trial—Raymond Santana and Kevin Richardson. After midnight their interrogators learned of the brutal rape of the investment banker. Santana, Richardson, and other teenagers who had been picked up by the police implicated Antron McCray, Kharey Wise, and Yusef Salaam in the roaming attacks. They were all contacted by the police the next day and voluntarily reported to them.[3]

The appalling savagery of the assault triggered off a frenzy among the local and national press, as newspapers vied to outdo each other with tabloid headlines and racially inflammatory rhetoric. In the meantime, Linda Fairstein, the head of the Manhattan District Attorney's Sex Crimes Prosecution Unit, who had already established a high public profile on the basis of her prosecutions of sex crimes, arranged to have the five defendants isolated. After more than twenty hours of relentless interrogation, four of the teenagers confessed to the crime of rape on videotape and in writing. The case was split into two separate trials in 1990 and all five of the defendants were found guilty of rape; other charges included sexual abuse, riot, assault, and robbery. The defendants were sentenced to prison terms ranging from five to fifteen years.

Developer Donald Trump set the tone for the atmosphere surrounding the case when he took out four full-page ads that ran in the New York Times, the Daily News, the New York Post, and Newsday on May 1, 1989, in which he called for the restoration of the death penalty and ranted against the "roving bands of wild criminals [that] roam our neighborhoods." "I want to hate these muggers and murderers," Trump proclaimed. "They should be forced to suffer and, when they kill, they should be executed for their crimes."[4] The mainstream press actively participated in creating the ugly mood: "Almost every member of the white-dominated press accepted without much question that mindless black and Latino adolescents could go from wreaking violent havoc in the park that night to carrying out a vicious gang rape."[5]

Widespread public revulsion about the crime and a rising tide of anger against the antisocial behavior of the teenagers apparently undercut the possibility of mobilizing local or national support for the accused—almost as if any public display of sympathy for their plight would detract from the heinous nature of the crime. And the mothers of the accused were also generally ignored, even though they lacked the necessary funds for proper legal representation: when they appeared in the press they were most often depicted as irresponsible or out-of-control ghetto mothers.

Nevertheless, there were some dissenting voices, particularly in Harlem, where the Amsterdam News issued clenched-teeth warnings that the teenagers were being transformed into the Scottsboro Boys and that a legal lynching was at hand.[6] The Reverend Calvin O. Butts, pastor of the Abyssinian Baptist Church, who led a delegation of fifteen black clergymen to the trial in a show of support for the mothers of the defendants, also invoked the Scottsboro case in his comments to the press.

In retrospect, the dissenters were proven to be closer to the truth. The 2002 confession of convicted serial rapist and murderer Matias Reyes to the brutal rape of the Central Park jogger, buttressed by convincing DNA evidence, brought the Scottsboro case back into sharp relief—and the parallels between the two cases were uncanny.

Unlike the accusers in the Scottsboro case, Victoria Price and Ruby Bates, the Central Park jogger came from the opposite side of the socioeconomic spectrum from the defendants—and she was literally in no position to identify her attacker or to participate in the narrative constructed about her. In the hands of the press, she became a perfect victim (even though her dramatic appearance at the trial and her composed, stoic bearing, free of self-pity, strongly suggested that this was not an image she would have chosen), and the police and the prosecutors were depicted as avenging angels on her behalf. Nevertheless, the basic elements of the cases were the same: young teenagers of color, from lower socioeconomic backgrounds, a rush to judgment by police and prosecutors, coerced confessions, false testimony, the absence of forensic evidence, and—most damning—the use of the judicial system as an instrument of revenge.[7]

In the aftermath of Matias Reyes's dramatic confession, Manhattan District Attorney Robert Morgenthau was required to review every detail of the case in the face of ferocious internal and public pressure to uphold the guilty verdicts against the five defendants. On December 5, 2002 Morgenthau argued before the State Supreme Court that all of the verdicts should be set aside, and the court agreed—leaving in its wake a decision that is still shaped by passionate and acrimonious debates about the guilt or innocence of the defendants, judicial and journalistic soul-searching, and sober, probing reflection about structural inequities in the American judicial system.

In the celebrated O. J. Simpson trial and the Duke University rape case, the Scottsboro case functioned more casually and referentially; that is to say, its invocation was less central in shaping public understandings of the cases. Abigail Thernstrom, a senior fellow at the Manhattan Institute in New York and Vice-Chair of the U.S. Commission on Civil Rights, pointed to poet Nikki Giovanni's remark in a September 1994 *Ebony* magazine symposium—"This is nothing but another Scottsboro case. . . . This kind of fascism cannot be allowed"[8]—and to defense attorney Johnnie Cochran's reference to the case in his closing remarks, as examples of how the Simpson trial could not be compared to Scottsboro by any stretch of the imagination. In the hands of Giovanni, Cochran, and others, Thernstrom argued, references to Scottsboro functioned as registers of long-standing racial grievances against the American justice system that threatened a high cost to the United States, "raising the level of despair and anger, and inviting just the sort of anger that was on display in the post-verdict days."[9] Thernstrom's argument carries with it an implicit view of history as well, suggesting as it does that such comparisons are ludicrous because the Simpson trial signifies how much social conditions have fundamentally changed in the United States since the 1930s.

Similarly, references to the Scottsboro case unexpectedly appeared after the celebrated 2006 Duke University rape case, in which three members of the university lacrosse team were accused of rape by one of two African American exotic dancers—a mother and student at neighboring North Carolina Central University—during an off-campus party at a house rented by members of the team. In the immediate aftermath of the event, Duke University suspended two upcoming games of the lacrosse team, forced the coach to resign, and canceled the rest of the season.

The subsequent furor, which was sustained through the successful reelection campaign of Durham District Attorney Mike Nifong, who publicly criticized members of the lacrosse team for failing to cooperate and vowed to vigorously pursue the case, continued for more than four hundred days. The rape accusation was proven to be false; District Attorney Nifong withdrew from the case in January 2007, and was subsequently disbarred by the North Carolina State Bar for his reckless and irresponsible prosecution of the case. In April 2007 North Carolina Attorney General Roy Cooper publicly dropped all charges against the three Duke lacrosse players—Collin Finnerty, Reade Seligmann, and David Evans.

With the growing public recognition that the three accused Duke University students had been victimized, losing at least a year of their lives and, undoubtedly, facing much longer-lasting effects, public recriminations continued.[10] The Scottsboro case was once again invoked. There were some striking parallels, to be sure, most notably the false accusation of rape (albeit by an African American woman), the absence of any forensic evidence, and the prosecutorial zeal of an unscrupulous District Attorney. In the eyes of some commentators, this apparent racial role-reversal signified the legacy of unresolved racial grievances: the Duke University rape case functioned as payback for racial injustices from long ago—such as the Scottsboro case. Thus, in an op-ed piece for the *Wall Street Journal*, John Steele Gordon went to great lengths to recapitulate the historic Scottsboro case in order to underscore his primary argument: that the Duke rape case represented "racial role reversal," a rush to judgment in which historical racial roles have—thankfully—been turned topsy-turvy and "the three Duke boys were guilty only of being white and affluent."[11] It is, he argued, the latest manifestation of how "political correctness" has distorted basic American values: "The country has come a long, long way in regard to race relations since 1931. But we have not yet reached the promised land where race is irrelevant. Far too many people are still being judged according to the color of their skin, not the content of their character, let alone the evidence."[12]

The End – We Hope

Figure 9.1 "The End—We Hope." Source: *Montgomery Advertiser*, October 28, 1976

Figure 9.2 "The Scottsboro Trial Rehash." Source: *Montgomery Advertiser*, July 12, 1977.

The Scottsboro Narrative emerged out of a specific set of historical circumstances, reflected a sensibility, a social and political ideology, a way of understanding the systemic issues of race and justice in the United States. Over the years it has worked its way into the underlying structure of an American vernacular language about race and justice. Its periodic resurfacing in the context of American racial spectacles strongly indicates that the Scottsboro case still fulfills important social and cultural functions in American life. Its meanings and associations may have changed somewhat over time, which is always the case with language and symbols, but the core of its references remain intact. Enshrined in public memory in this way, the Scottsboro case may at least have the virtue of assisting subsequent generations of Americans in unraveling the tangled web of race and injustice that continues to plague the United States. After all, what is the value of historical consciousness, embodied in American language, if it does not serve as a brake on platitudes and empty rhetoric, hasty judgments, and intemperate action?

Notes

IIIIIIIIIIIIIIIIIIIIIIIIIIIIIIIIIIIIIIIIIIIIIIIIIIIIIIIIIIIIIIIIIIIIIIIIIIIIIIIIIIIIIIIIIIIIIIIIIIIIIIIIIIIIIIIIIIIIIIIIIIIIIIIIIIIIIIIIIIIIIIIIIIIIIIIIIIIIIIIII

Introduction

1. John Lovell, "Review of Allan K. Chalmers's *They Shall Be Free*," *Journal of Negro Education* Winter 1952: 42.

2. Allen Ginsberg, "America," in *Howl and Other Poems* (San Francisco: City Lights Books, 1956).

3. "Activists Urge Prosecution in 1946 Killing of 4 Blacks," *Washington Post*, April 3, 2005, A14.

4. Courtland Milloy, "3 Howard Teens Missed Lessons Tested by Life," *Washington Post*, April 26, 2004, B01.

5. Ibid.

6. *Boston Globe*, April 23, 1997, A18.

7. Lori S. Robinson, "I Was Raped," *Emerge Magazine*, May 1997, 42–53.

8. Abigail Thernstrom and Henry D. Fetter, "From Scottsboro to Simpson," *Public Interest*, Winter 1996, 17–27.

9. "Fighting the Death Penalty: The Case of the Scottsboro Boys" (document in the possession of the author).

10. For a sampling of autobiographies and memoirs, see George Charney, *A Long Journey* (Chicago: Quadrangle Books, 1968); Benjamin J. Davis, *Communist Councilman from Harlem* (New York: International Publishers, 1969); Virginia Foster Durr, *Outside the Magic Circle: The Autobiography of Virginia Foster Durr*, ed. Hollinger F. Barnard (New York: Simon and Schuster, 1985); John Hammond and Irving Townsend, *John Hammond on Record: An Autobiography* (New York: Ridge Press, 1977); Harry Haywood, *Black Bolshevik: Autobiography of an Afro-American Communist* (Chicago: Liberator Press, 1978); Dorothy Healy and Maurice Isserman, *Dorothy Healey Remembers: A Life in the American Communist Party* (New York: Oxford University Press, 1990); Angelo Herndon, *Let Me Live* (New York: Arno Press/ New York Times, 1990); Nell Irvin Painter, *The Narrative of Hosea Hudson: His Life as a Communist in the South* (Cambridge, MA: Harvard University Press, 1979); Katherine Du Pre Lumpkin, *The Making of a Southerner* (New York: Alfred A. Knopf, 1947); Rosa Parks, with Jim Haskins, *My Story* (New York: Dial Press, 1992); William L. Patterson, *The Man Who Cried Genocide* (New York: International Publishers, 1971); Walter A. White, *A Man Called White* (Athens: University of Georgia Press, 1995); and Roy Wilkins, with Tom Mathews, *Standing Fast: The Autobiography of Roy Wilkins* (New York: Viking Press, 1982).

11. Two important early articles note this changing pattern: Hugh Murray, "Changing America and the Changing Image of Scottsboro," *Phylon* 38, no. 1 (1st Qtr., 1977): 82–92, and Carroll Van West, "Perpetuating the Myth of America: Scottsboro and Its Interpreters," *South Atlantic Quarterly* 80, no. 1 (Winter 1981): 36–48.

CHAPTER 1
FRAMING THE SCOTTSBORO BOYS

1. Patricia A. Schechter, *Ida B. Wells-Barnett and American Reform, 1880–1930* (Chapel Hill: University of North Carolina Press, 2001) 136–37.

2. Grace Elizabeth Hale, *Making Whiteness: The Culture of Segregation in the South, 1890–1940* (New York: Vintage, 1999), 204.

3. Ibid., 206.

4. *Chattanooga News*, March 26, 1931.

5. James S. Allen, "Organizing in the Deep South: A Communist's Memoir," *Nature, Society, and Thought* 13, no. 1 (January 2000): 41.

6. Helen Marcy, "Whip Up Lynch Mobs against 9 Negroes in Alabama," *Southern Worker*, April 4, 1931, 1–2.

7. Helen Marcy, "'Save Us' Negro Boys Write Folk in Chattanooga," *Southern Worker*, April 18, 1931, 1–2.

8. Ibid.

9. Mary Frederickson, "Interview with Hollace Ransdall," November 6, 1974 (Southern Oral History Project [#4007], Wilson Library, University of North Carolina at Chapel Hill).

10. Ibid., 10–11.

11. Hollace Ransdall, "Report on the Scottsboro, Ala, Case" (unpublished ms.), 8.

12. Ibid., 9.

13. Ibid., 9.

14. Ibid., 13.

15. Ibid., 13.

16. Ibid., 14.

17. Ibid., 14.

18. Ibid., 16.

19. Ibid., 18.

20. Ibid., 20.

21. See James A. Miller, Susan D. Pennybacker, and Eve Rosenhaft, "Mother Ada Wright and the International Campaign to Free the Scottsboro Boys, 1931–1934, *American Historical Review* 106, no. 2 (April 2001): 387–430. For the exchanges between Tom Johnson and the Comintern, see the Russian State Archives of Social and Political History (RGASPI) RGASPI 515/2285: 26; see also 28–30, 34b, 35–40, 59–60.

22. RGASPI 515/2285: 26; misspellings in the original.

23. RGASPI 515/2285: 26.

24. RGASPI 515/2285: 28.

25. RGASPI 515/2285: 29.

26. RGASPI 515/2285: 29.

27. W. L. Patterson, "Judge Lynch Goes to Court," *Labor Defender*, May 1931, 84, 99.

28. See Miller, Pennybacker, and Rosenhaft, "Mother Ada Wright."

29. Paul Peters, "From Darkest America," *Labor Defender*, February 1932, 31.

30. "Scottsboro Boys Write from Kilby Death Cells," *Labor Defender*, May 1932, 92.

31. See Miller, Pennybacker, and Rosenhaft, "Mother Ada Wright," 391.

32. James S. Allen,"The Scottsboro Struggle, *The Communist*, May 12, 1933, 440.

33. Dr. P. A. Stephens to Walter White, April 2, 1931 (NAACP Papers).

34. WTA to Dr. P. A. Stephens, April 9, 1931 (NAACP Papers).

35. Walter White to Dr. P. A. Stephens, April 20, 1931 (NAACP Papers).

36. NAACP Papers.

37. *New York Daily Worker*, April 24, 1931. See also Sheldon Avery, *Up from Washington: William Pickens and the Negro Struggle for Equality, 1900–1954* (Newark: University of Delaware Press, 2004), 122–23.

38. Ibid.

39. Walter White to Bob and Herb, May 3, 1931 (NAACP Papers).

40. For detailed accounts of the battle between the ILD and the NAACP, see James Goodman, *Stories of Scottsboro* (New York: Random House, 1995); Dan T. Carter, *Scottsboro: A Tragedy of the American South*, rev. ed. (Baton Rouge: Louisiana State University Press, 1979); Solomon, *The Cry Was Unity: Communists and African Americans, 1917–1936* (Jackson: University Press of Mississippi, 1998); Wilson Record, *Race and Radicalism: the NAACP and the Communist Party in Conflict* (Ithaca: Cornell University Press, 1964); and Hugh Murray, "The NAACP versus the Communist Party: The Scottsboro Rape Cases, 1931–1932," in Bernard Sternsher, ed., *The Negro in Depression and War: Prelude to Revolution, 1930–1945* (Chicago: Quadrangle Books, 1969).

41. Walter White to Lud, April 29, 1931 (NAACP Papers).

42. Walter White to Bob and Herb, May 3, 1931 (NAACP Papers).

43. Walter White to Olen Montgomery, May 14, 1931 (NAACP Papers).

44. Walter White to Dean (William Pickens), May 12, 1931 (NAACP Papers).

45. Walter White to Dean (William Pickens), May 27, 1931 (NAACP Papers).

46. Walter White to Dean (William Pickens), May 28, 1931 (NAACP Papers).

47. "A Talk by William Pickens to the 8 Negro Youths in Kilby Prison, Montgomery, Ala.," May 31, 1931 (NAACP Papers).

48. William Pickens to Walter White. May 31, 1931 (NAACP Papers).

49. "A Talk by William Pickens."

50. William Pickens to Walter White. June 6, 1931 (NAACP Papers).

51. "Confidential Excerpts from Report to National Office from Mr. William Pickens, Field Secretary, N.A.A.C.P. Re Scottsboro Cases," June 18, 1931 (NAACP Papers).

52. See Miller, Pennybacker, and Rosenhaft, "Mother Ada Wright."

53. Walter White, "The Negro and the Communists," *Harper's Monthly*, December 1931, 62–72; rpt. in Philip S. Foner and Herbert Shapiro, eds., *American Communism and Black Americans: A Documentary History, 1931–1934* (Philadelphia: Temple University Press, 1991), 274–88.

54. Memorandum re Telephone Conversation with Mr. Beddow—August 5, 1931 (NAACP Papers). Spellings of the names are per the original. Note that Haywood and Heywood have been used interchangeably throughout the literature.

55. Walter White to Claude Patterson, August 20, 1931 (NAACP Papers).

56. Ibid.

57. Walter White to Mr. Willie Robinson [*sic*], September 11, 1931 (NAACP Papers).

58. Walter White to Mr. Lincoln Steffens, November 10, 1931 (NAACP Papers).

59. Walter White to Mr. Ozie Powell, December 23, 1931, (NAACP Papers).

60. Scottsboro Defense Committee/International Labor Defense/League of Struggle for Negro Rights to Walter White, December 31, 1931 (NAACP Papers).

61. Ibid.

62. Ibid.

63. See Clarence Darrow, "Scottsboro," *Crisis*, March 1932, 81.

64. Walter White to Willie Robinson [*sic*]/Charley Weems, January 4, 1931 (i.e., 1932) (NAACP Papers).

65. White, "The Negro and the Communists."

66. Ibid., 277.

67. Ibid., 277.

68. Ibid., 277.

69. Ibid., 279.

70. Ibid., 281.

71. Ibid., 284.

72. Ibid., 288.

73. See Walter White, "The Scottsboro Cases" (for *The New Leader*), June 11, 1931 (NAACP Papers). Magdeleine Paz, "L'Affaire de Scottsboro," *Les Cahiers des Droits de l'Homme* 32 (new series), no. 21 (August 30, 1932): 491–96 (NAACP Papers).

74. Daniel Webster Hollis, *An Alabama Newspaper Tradition: Grover C. Hall and the Hall Family* (Tuscaloosa: University of Alabama Press, 1983).

75. Grover Hall, "Lay Down This Body of Death," *Montgomery Advertiser*, June 12, 1937.

76. J. Glenn Jordan, *The Unpublished Inside Story of the Infamous Scottsboro Case* (Huntsville, AL: White Printing Co., 1932), 2.

77. Ibid., 5.

78. Ibid., 5–6.

79. Ibid., 26.

80. Ibid., 32.

81. Ibid., 32.

82. Frank L. Owsley, "Scottsboro, the Third Crusade: The Sequel to Abolition and Reconstruction," *American Review* 1, no. 3 (June 1933): 257–85.

83. *I'll Take My Stand: The South and the Agrarian Tradition*, by 12 Southerners (New York: Harper and Row, 1962).

84. Frank Owsley, "The Irrepressible Conflict," in *I'll Take My Stand*, 61–91.

85. Owsley, "Scottsboro: The Third Crusade," 258.

86. Ibid., 258.

87. Ibid., 259.

88. Ibid., 261.

89. Ibid., 264, 266.

90. Ibid., 271, emphasis in the original.

91. Ibid., 271.

92. Ibid., 272.

93. Ibid., 273.

94. Ibid., 279.

95. Ibid., 282.

96. Ibid., 282, 283, 284.

97. Ibid., 285.

98. Ibid., 273.

99. Files Crenshaw, Jr. and Kenneth A. Miller, *Scottsboro: The Firebrand of Communism* (Montgomery, AL: Brown Printing Company, 1936).

100. Crenshaw and Miller, *Scottsboro*, Introduction.

101. Ibid., 54.

102. Ibid., 59–60.

103. Ibid.

104. Ibid., 62.

105. Ibid., 63.

106. Ibid., 65.

107. Ibid., 288.

108. Ibid., 72.

109. Ibid., 286.

110. Ibid., 291.

111. Ibid., 290.

112. Ibid., 297–98.

CHAPTER 2
"SCOTTSBORO, TOO": THE WRITER AS WITNESS

1. Countee Cullen, "Scottsboro, Too, Is Worth Its Song (A Poem to American Poets)," in *The Medea and Some Poems* (New York: Harper and Brothers, 1935).

2. Cary Nelson, *Repression and Recovery: Modern American Poetry and the Politics of Cultural Memory, 1919–1945* (Madison: University of Wisconsin Press, 1989), 207–8.

3. Countee Cullen, "Not Sacco and Vanzetti," in *The Black Christ and Other Poems* (New York: Harper & Brothers, 1929).

4. For a detailed discussion of Cullen's European travels, see Blanche E. Ferguson, *Countee Cullen and the Negro Renaissance* (New York: Dodd, Mead & Company, 1966).

5. Countee Cullen Papers, Library of Congress.

6. National Archives, Department of Justice, Mail and Files Division, 158260-46.

7. Quoted in David Levering Lewis, *When Harlem Was in Vogue* (New York: Alfred A. Knopf, 1984), 282.

8. Quoted in Earl Ofari Hutchinson, *Blacks and Reds: Race and Class in Conflict, 1919–1990* (East Lansing: Michigan State University Press, 1995), 141–42.

9. Gerald Early, ed., *My Soul's High Song: The Collected Writings of Countee Cullen, Voice of the Harlem Renaissance* (New York: Anchor Books, 1991), 3–4.

10. Countee Cullen, "Poet on Poet," *Opportunity*, February 1926, 73.

11. Arnold Rampersad, *The Life of Langston Hughes*, vol. I: *1902–1941: I, Too, Sing America* (New York: Oxford University Press, 1986), 216.

12. *The Collected Poems of Langston Hughes*, ed. Arnold Rampersad and David Roessel (New York: Alfred A. Knopf, 1994), 142–43.

13. Rampersad, *The Life of Langston Hughes*, I:219.

14. Langston Hughes, *Scottsboro Limited: Four Poems and a Play in Verse* (New York: Golden Stair Press, 1932).

15. See Langston Hughes, *I Wonder As I Wander: An Autobiographical Journey* (New York: Thunder's Mouth Press, 1989), 43–44; Faith Berry, *Before and Beyond Harlem: A Biography of Langston Hughes* (New York: Wings Books, 1995), 134–35; Rampersad, *The Life of Langston Hughes*, I:223.

16. Langston Hughes, "Cowards from the Colleges," in *Good Morning, Revolution: Uncollected Writings of Social Protest by Langston Hughes*, ed. Faith Berry (New York: Lawrence Hill & Company, 1973), 57.

17. Ibid., 57–58.

18. See Berry, *Before and Beyond Harlem*, 135–36, and Rampersad, *The Life of Langston Hughes*, I:224–25.

19. Langston Hughes, "Southern Gentlemen, White Prostitutes, Mill Owners, and Negroes," in *Good Morning, Revolution*, 49.

20. *Good Morning, Revolution*, 49.

21. *The Collected Poems of Langston Hughes*, 143.

22. Hughes, *I Wonder As I Wander*, 46.

23. Anthony Buttitta, "A Note on *Contempo* and Langston Hughes," *Negro: Anthology Made by Nancy Cunard, 1931–1933* (London, 1934; rpt. New York: Negro Universities Press, 1969), 142.

24. *The Collected Poems of Langston Hughes*, 31.

25. Ibid., 168.

26. See Granville Hicks, Joseph North, Michael Gold, et al., eds., *Proletarian Literature in the United States: An Anthology* (New York: International Publishers, 1935), 262. See also James Smethurst's discussion of *Scottsboro Limited* in *The New Red Negro: The Literary Left and African American Poetry, 1930–1946* (New York: Oxford University Press, 1999), 103–7.

27. See Mark Solomon, *The Cry Was Unity: Communists and African Americans, 1917–1936* (Jackson: University Press of Mississippi, 1998), 139–42. See also Mark Naison, *Communists in Harlem during the Depression* (New York: Grove Press, 1983), 447–49.

28. "Give Scottsboro Play Tonight," *Daily Worker*, n.d. (International Labor Defense Papers).

29. "Scottsboro," *Workers' Theatre*, April 1932, 14–17.

30. Ibid., 14.

31. Ibid., 17.

32. Ibid., 17.

33. Langston Hughes, *Scottsboro Limited*.

34. Smethurst, *The New Red Negro*, 104–5.

35. Hughes, *Scottsboro Limited*.

36. Ibid.

37. Ibid.

38. Ibid.

39. Hughes, *I Wonder As I Wander*, 61.

40. Langston Hughes, "Brown America in Jail: Kilby," *Opportunity*, June 1932, rpt. in *Good Morning Revolution*, 50.

41. Ibid., 61–62.

42. Quoted in Rampersad, *Life of Langston Hughes*, I:638. The date was of course August 19, not August 18.

43. *The Collected Poems of Langston Hughes*, 204.

44. Ibid., 206.

45. Herman J. D. Carter. *The Scottsboro Blues* (Nashville, TN: Mahlon Publishing Co., 1933), 3.

46. Ibid., 7.

47. Ibid., 9.

48. Ibid., 9.

49. Ibid., 10.

50. Ibid., 10–11.

51. Ibid., 14.

52. Louise Kertesz, *The Poetic Vision of Muriel Rukeyser* (Baton Rouge: Louisiana State University Press, 1980), 90.

53. Ibid., 22.

54. Muriel Rukeyser, "The Lynchings of Jesus," in *The Collected Poems* (New York: McGraw-Hill, 1978), 24. See Solomon, *The Cry Was Unity*, 19.

55. Rukeyser, "The Lynchings of Jesus," 25.

56. Ibid., 25.

57. Ibid., 25.

58. Ibid., 26.

59. Ibid., 26.

60. Ibid., 26.

61. Ibid., 26.

62. Ibid., 26.

63. Ibid., 27.

64. See Miller, Pennybacker, and Rosenhaft, "Mother Ada Wright."

65. *Workers Voice* (Chicago), October 1, 1932 (International Labor Defense Papers).

66. Nancy Mitford, *"The Pursuit of Love" and "Love in a Cold Climate"* (New York: Modern Library, 1994), 145. I am indebted to John Lonsdale for directing my attention to this work.

67. See *Nancy Cunard: Essays on Race and Empire*, ed. Maureen Moynagh (Orchard Park, NY: Broadview Press, Ltd., 2002). See also Anne Chisholm, *Nancy Cunard: A Biography* (New York: Knopf, 1979); Henry Crowder, *As Wonderful As All That? Henry Crowder's Memoir of His Affair with Nancy Cunard*, ed. Robert Allen (Navarro, CA: Wild Trees Press, 1987); Hugh Ford, ed., *Nancy Cunard: Brave Poet, Indomitable Rebel, 1896–1965* (Philadelphia: Chilton, 1968); and Jane Marcus, "Bonding and Bondage: Nancy Cunard and the Making of the Negro Anthology," in *Borders, Boundaries and Traces*, ed. Mae Gwendolyn Henderson (New York: Routledge, 1995), 33–63.

68. "Girl Who Hates 'So Many White People,'" *Daily Express*, July 6, 1933.

69. Ibid.

70. Cunard, *Negro*, iii.

71. Ibid., iii.

72. Ibid., 247.

73. Ibid., 264.

74. Sandra Whipple Spanier, *Kay Boyle, Artist and Activist* (Carbondale: Southern Illinois University Press, 1986); Joan Mellen, *Kay Boyle: Author of Herself* (New York: Farrar, Straus & Giroux, 1994). See also Shari Benstock, *Women of the Left Bank Paris, 1900–1940* (Austin: University of Texas Press, 1986).

75. Carleton Beals, "The Scottsboro Puppet Show," *The Nation*, February 5, 1936, 149–50; Beals, "Scottsboro Interview," *The Nation*, February 12, 1936, 178–79.

76. Kay Boyle, "A Communication to Nancy Cunard," in *Collected Poems of Kay Boyle* (Port Townsend, WA: Copper Canyon Press, 1991), 66.

77. Ellen McWhorter, "A Note on 'Communication to Nancy Cunard'" (www .english.uiuc.edu/maps/poets/a_f/boyle/cunard.htm).

78. Robert Hayden, "These Are My People (A Mass Chant)," *Heart-Shape in the Dust*, (Detroit: Falcon Press, 1940), 61–62.

CHAPTER 3
STAGING SCOTTSBORO

1. Malcolm Goldstein, *The Political Stage: American Drama and Theater of the Great Depression* (New York: Oxford University Press, 1974), 38. See also *Workers Theatre*, May–June 1938, 18.

2. Morgan Yale Himelstein, *Drama Was a Weapon: The Left-Wing Theatre in New York, 1929–1941* (New Brunswick, NJ: Rutgers University Press, 1963).

3. John Chapman, "Legal Murder," *New York Daily News*, February 17, 1934.

4. "Scottsboro" (International Labor Defense Papers).

5. W. A. Ucker, "'Legal Murder,'" Play Based on Scottsboro Case, Opens at President Theatre" (International Labor Defense Papers).

6. Ibid.

7. Patrick McGilligan and Paul Buhle, *Tender Comrades: A Backstory of the Hollywood Blacklist* (New York: St. Martin's Press, 1997), 707.

8. John Wexley, *They Shall Not Die* (New York: Alfred A. Knopf, 1934), 19–20.

9. Ibid., 20.

10. Ibid., 42–44.

11. Ibid., 66.

12. Ibid., 86.

13. Ibid., 96.

14. Ibid., 103.

15. Ibid., 108.

16. Ibid., 138–39.

17. Ibid., 168–70.

18. Ibid., 181.

19. Ibid., 190.

20. Ibid., 191.

21. F. Raymond Daniell, "An Alabama Court in Forty-fifth Street," *New York Times*, March 4, 1934, sec. 9, 1.

22. John Anderson, "'They Shall Not Die,'" *New York Evening Journal*, February 22, 1934, 12.

23. "The Theater by Arthur Ruhl," *New York Herald Tribune*, February 22, 1934, 16.

24. Ibid.

25. Ibid.

26. Robert Garland, "Righteous Indignation Surges in the Royale," *New York World Telegram*, February 27, 1934, 32.

27. Ibid.

28. Isidor Schneider, "They Shall Not Die," *Labor Defender*, May 1934, 35.

29. Ibid.

30. Ibid.

31. Harold Edgar, "They Shall Not Die," *Daily Worker*, February 23, 1934, 7.

32. Ibid.

33. Ibid.

34. Michael Gold, "Change the World," *Daily Worker*, April 11, 1934, 5.

35. "Scottsboro Play Symposium Issue," *New York Times*, March 5, 1934.

36. Ibid.

37. Ibid.

38. Letter from George Leon Hughes of 274 W. 140th Street, New York City, cited in Crenshaw and Miller, *Scottsboro*, 291–92.

39. Himelstein, *Drama Was a Weapon*, 50.

40. Quoted in Goldstein, *The Political Stage*, 59.

41. Ibid., 66.

42. Sender Garlin, "Paul Peters—Revolutionary Playwright—An Interview," *Daily Worker*, May 15, 1934, 5.

43. Paul Peters and George Sklar, *Stevedore* (New York: Covici-Friede, 1934), 14–15.

44. Ibid., 24.

45. Ibid., 40.

46. Ibid., 53.

47. Ibid., 61.

48. Ibid., 61–62.

49. Ibid., 106–7.

50. Ibid., 112.

51. Ibid., 121.

52. See Goldstein, *The Political Stage*, 67.

53. See Nancy Wick, "Playing with History: The Revival of a 1936 Black Drama—Part of a Controversial New Deal Project—Fulfills the Dream of One Determined UW Director" (www.washington.edu/alumni/columns/dec95/stevedore .html).

54. Joseph Wood Krutch, "Drama," *Nation*, May 2, 1934, 516.

55. Ibid., 516.

56. Michael Gold, "Stevedore," *New Masses*, May 1, 1934, 28.

57. John Howard Lawson, "Straight from the Shoulder" *New Theatre*, November 1934, 12.

58. Liston Oak, "Theatre Union Replies," *New Theatre*, November 1934, 12.

59. Clarence Muse to Walter White, April 20, 1935 (NAACP Papers).

60. Ibid.

61. Ibid.

62. Assistant Secretary (Roy Wilkins) to Mr. Claude A. Barnett, June 19, 1933 (NAACP Papers).

63. Floyd C. Covington to Mr. Walter White, July 31, 1933 (NAACP Papers).

64. F.S.N., "America's Juvenile Hoboes," *New York Times*, September 22, 1933, Amusements, 14.

65. Patrick McGilligan, *Fritz Lang: The Nature of the Beast* (New York: St. Martin's Press, 1997), 223.

66. Arthur Just Hartley to Mr. William Jay Schieffelin, June 15, 1936 (NAACP Papers).

67. McGilligan, *Fritz Lang*, 227. See in particular Peter Bogdanovich, *Fritz Lang in America* (New York: Praeger, 1969).

68. McGilligan., *Fritz Lang*, 227.

69. Ibid., 227.

CHAPTER 4
FICTIONAL SCOTTSBOROS

1. Suzanne Sowinska, Introduction to Grace Lumpkin, *To Make My Bread* (Urbana: University of Illinois Press, 1995), x. For the fullest and richest account of Grace Lumpkin to date, see Jacquelyn Dowd Hall, "Women Writers, the 'Southern Front,' and the Dialectical Imagination," *Journal of Southern History* 69, no. 1 (February 2003): 3–38. James A. Miller, "The Limits of Radical Reform: Grace Lumpkin and the Fate of Proletarian Fiction," unpublished ms.

2. Grace Lumpkin, *A Sign for Cain* (New York: Lee Furman, Inc., 1935), 46.

3. Ibid., 52–53.

4. Ibid., 202.

5. Ibid., 202.

6. Ibid., 233–34.

7. Ibid., 245.

8. Ibid., 268.

9. Ibid., 264.

10. Ibid., 263.

11. Ibid., 306.

12. Ibid., 311–12.

13. Ibid., 371–72.

14. Ibid., 374.

15. *Boston Transcript*, October 26, 1935, 5.

16. *Books*, October 27, 1935, 4.

17. *New Republic*, October 23, 1935.

18. *New York Times*, December 15, 1935, 7.

19. See Sowinska, Introduction to Lumpkin, *To Make My Bread*, xvii. See in particular Hall, "Women Writers."

20. Alan Wald, "The Subaltern Speaks," in *Writings from the Left: New Essays in Radical Culture and Politics* (New York: Verso, 1994), 179–80. See also Alan Wald, *Exiles from a Future Time: The Forging of the Mid-Twentieth-Century American Left* (Chapel Hill: University of North Carolina Press, 2002).

21. Wald, *Exiles from a Future Time*, 1.

22. Guy Endore, *The Crime at Scottsboro* (Hollywood, CA: Hollywood Scottsboro Committee, 1938).

23. See in particular Barbara Foley's discussion of *Babouk* in *Radical Representations: Politics and Form in U.S. Proletarian Fiction, 1929–1941* (Durham, NC: Duke University Press, 1993).

24. Guy Endore, *Babouk* (New York: Monthly Review Press, 1991), 51–52.

25. Ibid., 52–53.

26. Ibid., 53.

27. Ibid., 102.

28. Ibid., 96–97.

29. Ibid., 157.

30. Ibid., 168.

31. Ibid., 175.

32. Ibid., 182.

33. John Chamberlain, "Books of the Times," *New York Times*, August 31, 1934.

34. Paul Allen, *Books*, September 23, 1934, 20.

35. Martha Gruening, in *New Republic*, October 17, 1934.

36. Eugene Gordon, "Black and White, Unite and Fight," *New Masses*, October 23, 1934, 24.

37. Wald, "The Subaltern Speaks," 186.

38. Arnold Rampersad, "Introduction to the 1992 Edition," Arna Bontemps, *Black Thunder* (Boston: Beacon Press, 1992), x. Cile Moise Traywick, "Coming to Terms with the Scottsboro Trials: Arna Bontemps and *Black Thunder*" (unpublished ms., English 757, University of South Carolina–Columbia, June 6, 1997).

39. Arna Bontemps, "Why I Returned (A Personal Essay)," in *The Old South* (New York: Dodd, Mead & Company, 1973), 13–15.

40. Arna Bontemps, "Introduction to the 1968 Edition," *Black Thunder*, xxviii.

41. Bontemps, *Black Thunder*, 63.

42. Ibid., 20.

43. Ibid., 21.

44. Ibid., 21.

45. Ibid., 24.

46. Ibid., 76.

47. Ibid., 152.

48. Ibid., 89.

49. Ibid., 90, 91.

50. Ibid., 106–7.

51. Ibid., 168–69.

52. Ibid., 203–4.

53. Ibid., 56.

54. Ibid., 62.

55. Ibid., 143.

56. Ibid., 168.

57. Gruening, "Review," *New Republic*, February 26, 1936, 91.

58. J. D., "Fiction," *Saturday Review of Literature*, February 15, 1936, 27.

59. Lucy Tomkins, "Slaves' Rebellion," *New York Times Book Review*, February 2, 1936, 7.

CHAPTER 5
RICHARD WRIGHT'S SCOTTSBORO OF THE IMAGINATION

1. *Lawd Today!* in Richard Wright, *Early Works*, ed. Arnold Rampersad (New York: Library of America, 1991), 176.

2. *Native Son*, in Wright, *Early Works*, 495.

3. Ibid., 515.

4. "How Bigger Was Born," in Wright, *Early Works*, 874.

5. Quoted in Keneth Kinnamon, *The Emergence of Richard Wright* (Urbana: University of Illinois Press, 1972), 51.

6. Richard Wright, "I Tried to Be a Communist," in Richard Crossman, ed., *The God That Failed* (New York: Harper, 1950).

7. Hazel Rowley, *Richard Wright: The Life and Times* (New York: Henry Holt and Company, 2001), 78.

8. Richard Wright, "Blueprint for Negro Writing," *New Challenge*, Fall 1937, 53–65.

9. Michel Fabre, *The Unfinished Quest of Richard Wright*, 2nd ed. (Urbana: University of Illinois Press, 1993).

10. Richard Wright, "Between the World and Me," *Partisan Review* 32 (July–August 1935); rpt. in Ellen Wright and Michel Fabre, eds., *Richard Wright Reader* (New York: Harper and Row, 1978), p.246–47.

11. Wright, "I Tried to Be a Communist."

12. Richard Wright, "Big Boy Leaves Home," In *Early Works*, 255.

13. Ibid., 270–71.

14. Richard Wright, *Black Boy (American Hunger)*, in Wright, *Later Works*, 324–25.

15. Richard Wright, "Fire and Cloud," in Wright, *Early Works*, 357.

16. Ibid., 368–69.

17. Ibid., 376–77.

18. Ibid., 382.

19. Ibid., 397–98.

20. Wright, "Blueprint for Negro Writing," 63.

21. "Fire and Cloud," 406.

22. For a discussion of the genesis of *Uncle Tom's Children*, see Fabre, *The Unfinished Quest of Richard Wright*, 156–65.

23. Richard Wright, "Bright and Morning Star," in Wright, *Early Works*, 409.

24. Ibid., 410.

25. Ibid., 410.

26. Ibid., 416–18.

27. Kinnamon, *The Emergence of Richard Wright*, 114.

28. Ibid., 414.

29. Ibid., 422.

30. Ibid., 433.

31. Ibid., 433.

32. Ibid., 441.

33. Ibid., 435.

34. See Robin Kelley, *Hammer and Hoe: Alabama Communists during the Great Depression* (Chapel Hill: University of North Carolina Press, 1990), and James S. Allen,

"Organizing in the Deep South: A Communist's Memoir," *Nature, Science and Thought* 3 (January 2001).

35. Wright, *Native Son*, 524.

36. Arnold Rampersad forcefully argues that the changes mandated by the Book-of-the-Month Club "almost emasculated Bigger Thomas." See "Too Honest for His Own Time," in *The Critical Response to Richard Wright*, ed. Robert J. Butler (Westport, CT: Greenwood Press, 1995).

37. For detailed accounts of the Nixon case, see Kinnamon, *The Emergence of Richard Wright*, and Fabre, *The Unfinished Quest of Richard Wright*. See also Rowley, *Richard Wright*.

38. Wright, *Native Son*, 475.

39. Fabre, *The Unfinished Quest of Richard Wright*, 170.

40. See Rowley, *Richard Wright*, 94–95.

41. See Constance Webb, *Richard Wright: A Biography* (New York: Putnam's, 1968).

42. Wright, *Native Son*, 714.

43. See, for example, William Maxwell's discussion of *Native Son* in *New Negro, Old Left: African-American Writing and Communism between the Wars* (New York: Columbia University Press, 1999).

44. Wright, *Native Son*, 770.

45. Ibid., 796.

46. Benjamin Davis, Jr., "Review of *Native Son*," *Sunday Worker*, April 14, 1940, 4, 6.

47. Wright, *Native Son*, 824.

48. Ibid., 804.

49. Ibid., 809.

50. Ibid., 818.

51. Ibid., 820.

52. Ibid., 821.

53. Ibid., 823.

54. Ibid., 825.

55. Ibid., 849.

56. Ibid., 849.

57. Richard Wright, *Eight Men* (New York: HarperCollins, 1996). This collection of short stories was first published posthumously in 1961 (Cleveland: World Publishing).

58. Mary Helen Washington, "The Phylon Symposium: 1950, and Other Signs of Black Cold War Anxiety" (unpublished ms).

59. William Gardner Smith, *Last of the Conquerors* (New York: Farrar, Straus, 1948), 107.

60. William Demby, *Bettlecreek* (Jackson: University Press of Mississippi, 1998), 151.

CHAPTER 6
THE SCOTTSBORO DEFENDANT AS PROTO-REVOLUTIONARY: HAYWOOD PATTERSON

1. Haywood Patterson and Earl Conrad, *Scottsboro Boy* (Garden City, NY: Doubleday & Co., Inc., 1950), 18.

2. Kwando Mbiassi Kinshasa, *The Man from Scottsboro: Clarence Norris in His Own Words* (Jefferson, NC: McFarland & Company, Inc., 1997), 70.

3. Carleton Beals, "Scottsboro Interview," *The Nation*, February 12, 1936, 178–79.

4. Patterson and Conrad, *Scottsboro Boy*, 4–5.

5. Ibid., 74.

6. I. F. Stone, "Fugitive's Book Tells of Nazi-Like Horror Camps in Southland, U.S.A.," *Gazette and Daily*, York, PA, June 2, 1950.

7. Ibid.

8. Ibid.

9. Ibid.

10. Ibid.

11. See Robert B. Stepto, *From Behind the Veil: A Study of Afro-American Narrative* (Urbana: University of Illinois Press, 1979); William L. Andrews, *To Tell a Free Story: The First Century of Afro-American Autobiography* (Urbana: University of Illinois Press, 1986); Valerie Smith, *Self-Discovery and Authority in Afro-American Narrative* (Cambridge: Harvard University Press, 1987); Kenneth Mostern, *Autobiography and Black Identity Politics: Racialization in Twentieth-Century America* (New York: Cambridge University Press, 1999); and Harryette Mullen, *Freeing the Soul: Race, Subjectivity, and Differences in Slave Narratives* (New York: Cambridge University Press, 1999). See also H. Bruce Franklin, *The Victim As Criminal and Artist* (New York: Oxford University Press, 1978), and *Prison Literature in America: The Victim As Criminal and Artist*, expanded ed. (New York: Oxford University Press, 1989).

12. See Stepto, *From Behind the Veil*.

13. It is striking that H. Bruce Franklin makes only a passing reference to this seminal work.

14. Patterson and Conrad, *Scottsboro Boy*, 22–23.

15. Ibid., 41–42.

16. Ibid., 31.

17. Ibid., 31.

18. Haywood Patterson to Mr. J. Louis Engdahl, Dec.10, 1931 (International Labor Defense Papers).

19. Haywood Patterson to William L. Patterson, December 19, 1932 (International Labor Defense Papers).

20. Haywood Patterson to Mr. John Wexley, March 5, 1934 (International Labor Defense Papers).

21. Patterson and Conrad, *Scottsboro Boy*, 47–48.

22. Haywood Patterson to Morris Shapiro, April 11, 1936 (NAACP Papers).

23. Haywood Patterson to Juanita E. Jackson, December 11, 1936 (NAACP Papers).

24. Ibid.

25. Haywood Patterson to Morris Shapiro, January 2, 1937 (NAACP Papers).

26. Haywood Patterson to Anna Damon, August 22, 1936 (International Labor Defense Papers).

27. Haywood Patterson to Anna Damon, October 28, 1936 (International Labor Defense Papers).

28. Haywood Patterson to Anna Damon, December 11, 1936 (International Labor Defense Papers).

29. Haywood Patterson to Anna Damon, January 11, 1937 (International Labor Defense Papers).

30. Ibid.

31. Ibid.

32. Anna Damon to Haywood Patterson, January 15, 1937 (International Labor Defense Papers).

33. Haywood Patterson to Anna Damon, January 18, 1937 (International Labor Defense Papers).

34. Ibid.

35. Ibid.

36. Rose Baron to Haywood Patterson, February 1, 1937 (International Labor Defense Papers).

37. Haywood Patterson to Rose Baron, February 5, 1937 (International Labor Defense Papers).

38. Haywood Patterson to Mrs. Hester Huntington, n.d. [1944] (International Labor Defense Papers).

39. Haywood Patterson to Bobby Boyle, October 20, 1937 (International Labor Defense Papers). Some of Patterson's correspondence to Bobby Boyle was faithfully reproduced in Kay Boyle's "A Communication to Nancy Cunard."

40. Haywood Patterson to Miss Maria Sulek, n.d. (International Labor Defense Papers).

41. Dick Cardamone to Anna Damon, Secy, Intl Labor Defense, June 28, 1943 (International Labor Defense Papers).

42. Don Jonson to Mrs. Huntington, September 29, 1945 (International Labor Defense Papers).

43. Louis Colman to Don Jonson, October 9, 1945 (International Labor Defense Papers).

44. Allan Chalmers to Mrs. Huntington, January 14, 1944 (International Labor Defense Papers).

45. Patterson and Conrad, *Scottsboro Boy*, 7.

46. Ibid., 15.

47. Ibid., 45.

48. Ibid., 45.

49. Ibid., 46.

50. Ibid., 47.

51. Ibid., 132–33.

52. Olen Montgomery to Anna Damon, July 18, 1936 (International Labor Defense Papers).

53. Patterson and Conrad, *Scottsboro Boy*, 79–80.

54. Ibid., 110.

55. Ibid., 98.

56. Ibid., 104.

57. Ibid., 191–92.

58. Patterson and Conrad, *Scottsboro Boy*, 244.

59. Ibid., 245–46.

60. Stone, "Fugitive's Book Tells of Nazi-Like Horror Camps."

61. Abner Berry, "American Dachau," *Masses and Mainstream*," July 1950, 83–86.

62. J. N. Popham, "Review of *Scottsboro Boy*," *New York Times*, June 4, 1950, 14.

63. Hodding Carter, "Cause Celebre and Cause CP," *Saturday Review of Literature*, June 10, 1950.

64. "Letter of Walter White," *Saturday Review of Literature*, July 22, 1950, 23.

65. Allan K. Chalmers, *They Shall Be Free* (Garden City, NY: Doubleday & Company, 1951), 222.

66. William A. Nolan, *Communism versus the Negro* (Chicago: Institute of Social Order, 1950), 18–19.

67. See James Goodman, *Stories of Scottsboro*, 380–81.

CHAPTER 7
COLD WAR SCOTTSBOROS

1. Michael Denning, *The Cultural Front: The Laboring of American Culture in the Twentieth Century* (New York: Verso, 1996), xvi.

2. Don M. Mankiewicz, *Trial* (New York: Harper & Brothers, 1955), 2.

3. Ibid., 8–9.

4. Ibid., 17.

5. Ibid., 30–31.

6. Ibid., 102.

7. Ibid., 104.

8. Samuel A. Stouffer, *Communism, Conformity, and Civil Liberties* (Garden City, NY: Doubleday & Company, 1955). See also Michael Barson, *Better Dead than Red!* (New York: Hyperion, 1992).

9. Crossman, ed., *The God That Failed*.

10. Mankiewicz, *Trial*. 197.

11. Ibid., 200.

12. Ibid., 36.

13. Ibid., 94–95.

14. Ibid., 141–42.

15. Ibid., 147–48.

16. Ibid., 261.

17. Ibid., 137–38.

18. Ibid., 222–23.

19. Ibid., 126.

20. Ibid., 136.

21. Ibid., 285.

22. Ibid., 304.

23. *Catholic World*, March 1955, 471.

24. *Times Literary Supplement*, July 15, 1955, 393.

25. Nora Sayre, *Running Time: Films of the Cold War* (New York: Dial Press, 1982), 80.

26. Ibid., 80–81.

27. Ibid., 81.

28. For an excellent discussion of these tensions in Lumpkin's life and work, see Wes Mantooth, "'You Factory Folks Who Sing This Song Will Surely Understand':

Cultural Representations in the Gastonia Novels of Myra Page, Grace Lumpkin, and Olive Dargan" (Ph.D. dissertation, Department of English, George Washington University, 2004).

29. *Book Review Digest*, 274.

30. See Sowinska, Introduction to Lumpkin, *To Make My Bread*, xiii.

31. Ibid.

32. Lorine Pruette, "Review of *The Wedding*, by Grace Lumpkin," *New York Herald Tribune Lively Arts and Book Review*, March 5, 1939, 10.

33. Jessica Kimble Printz, "Tracing the Fault Lines of the Radical Female Subject: Grace Lumpkin's *The Wedding*, in *Radical Revisions: Rereading 1930s Culture*, ed. Bill Mullen and Sherry Linkon (Urbana: University of Illinois Press, 1996), 167–86.

34. Ibid., 183.

35. Grace Lumpkin, *The Wedding* (New York: Lee Furman, Inc., 1939), 281–82.

36. See Sowinska, Introduction to Lumpkin, *To Make My Bread*. See also Frank N. D. Buchman, *Remaking the World* (New York: Robert M. McBride, 1949).

37. Sowinska, "Introduction."

38. Whittaker Chambers, *Witness* (New York: Random House, 1952).

39. Sowinska, "Introduction."

40. Grace Lumpkin, "The Law and the Spirit," *National Review*, May 25, 1957, 498–99. See also Lumpkin, "How I Returned to the Christian Faith," *Christian Economics*, June 23, 1964, 4.

41. "Jane Speed Gets Fine, 6 Months," *Montgomery Advertiser*, May 20, 1933, 1, 9.

42. "Rabbi Goldstein's Departure," *Montgomery Advertiser*, May 18, 1933.

43. Grace Lumpkin, *Full Circle* (Boston: Western Islands, 1962), 113–14.

44. Ibid., 114.

45. Ibid., 114–15.

46. Ibid., 117.

47. Ibid.

48. Ibid., 153.

49. Ibid., 153–54.

50. Ibid., 154.

51. Ibid., 155–56.

52. Ibid., 252.

53. Ibid., 289.

54. Ibid., 304.

CHAPTER 8
HARPER LEE'S *TO KILL A MOCKINGBIRD*

1. Harper Lee, *To Kill a Mockingbird* (New York: J.B. Lippincott, 1960; Warner Books Edition), 220. All further page references are to this latter edition.

2. See, for example, Kelly Covin's 1970 novel *Hear That Train Blow!*; NBC's 1976 television docudrama *Judge Horton and the Scottsboro Boys*; the 1996 film *A Time to Kill*, starring Matthew McConaughey, Sandra Bullock, and Samuel L. Jackson; and the Emmy Award–winning 2001 PBS documentary *Scottsboro: An American Tragedy*. In 2004, the *Chattanooga Times* announced that a production crew was in town shoot-

ing location shots for a feature film, "Heavens Fall," based on the Scottsboro case, starring Timothy Bottoms. The filmed was released in 2006.

3. See Claudia Durst Johnson, Understanding "To Kill a Mockingbird" (Westport, CT: Greenwood Press, 2002).

4. Phoebe Adams, "Review of To Kill a Mockingbird," Atlantic, August 1960, 98.

5. Jane Kansas, "To Kill a Mockingbird," www.chebucto.ca/culture/mockingbird.

6. Johnson, Understanding "To Kill a Mockingbird."

7. Ibid.

8. "Oscar Winner Gregory Peck Dies at 87," CNN.com, Entertainment, June 13, 2003, www.cnn.com/2003/SHOWBIZ/Movies/06/12/obit.peck/.

9. Eric J. Sundquist, "Blues for Atticus Finch: Scottsboro, Brown, and Harper Lee," in The South as an American Problem, ed. by Larry J. Griffin and Don H. Doyle (Athens: University of Georgia Press, 1995), 182, 183.

10. Ibid., 186.

11. "Publishing History of To Kill a Mockingbird," www.chebucto.ca/Culture/Mockingbird. See also Johnson, "Understanding To Kill a Mockingbird."

12. To Kill a Mockingbird, 116.

13. Ibid., 146–47.

14. Ibid., 5.

15. Ibid., 136.

16. Ibid., 30.

17. Seckatary Hawkins, http://seckatary.com/page2.html.

18. Sundquist, "Blues for Atticus Finch," 185. See also Arthur Raper, The Tragedy of Lynching (Chapel Hill, NC: University of North Carolina Press, 1933).

19. In his portrait of Harper Lee in Mockingbird, Charles J. Shields argues that "the scope of those trials was too big, too excessive for Nelle's purposes....huge in its implications and too far removed from Nelle's experiences." I clearly disagree. Shields, Mockingbird: A Portrait of Harper Lee (New York: Henry Holt and Company, 2006), 117–18.

20. To Kill a Mockingbird, 88.

21. Ibid., 75.

22. Ibid., 76.

23. Ibid., 76.

24. See James Goodman's discussion of Judge Horton in Stories of Scottsboro.

25. To Kill a Mockingbird, 157.

26. Ibid., 170.

27. Ibid., 179.

28. Ibid., 203.

29. Ibid., 192.

30. Ibid., 196.

31. Ibid., 216.

32. Ibid., 220.

33. Ibid., 220.

34. Ibid., 220.

35. Ibid., 221.

36. Ibid., 226.

37. Ibid., 226.
38. Ibid., 227.
39. Ibid., 216.
40. Ibid., 229.
41. Ibid., 234.
42. Ibid., 236.
43. Ibid., 241.
44. Byron Woodfin, *Lay Down with Dogs: The Story of Hugh Otis Bynum and the Scottsboro First Monday Bombing* (Tuscaloosa: University of Alabama Press, 1997), 14.
45. Ibid., 14.
46. See Dean Shackleford, "The Female Voice in *To Kill a Mockingbird*: Narrative Strategies in Film and Novel," *Mississippi Quarterly* 50, no. 1 (Winter 1996/97): 101–14.
47. Ibid., 102.
48. Jane Kansas, "To Kill a Mockingbird," www.chebucto.ca/culture/mockingbird.
49. Donald Bogle, *Toms, Coons, Mulattoes, Mammies, and Bucks: An Interpretive History of Blacks in American Films* (New York: Continuum Publishing Company, 1989), 203.
50. Ibid., 207.
51. *To Kill a Mockingbird*, 276.
52. Ibid., 276.

Epilogue

1. Sydney H. Schanberg, "A Journey through the Tangled Case of the Central Park Jogger," *Village Voice*, November 19, 2002.
2. Executive Summary—Central Park Jogger Case Panel Report, www.nyc.gov/html/nypd/html/dcpi/executivesumm_cpjc.html.
3. Ibid.
4. Lisa W. Foderaro, "Angered by Attack, Trump Urges Return of the Death Penalty," *New York Times*, May 1, 1989. See also Jimmy Breslin, "Trump's Ad Fueled the Fire: Central Park Jogger Case," *Newsday*, October 25, 2002.
5. Lynnell Hancock, "Wolf Pack: The Press and the Central Park Jogger," *Columbia Journalism Review* 41, no. 5 (January/February 2003): 38–42.
6. William Glaberson, "In Jogger Case, Once Viewed Starkly, Some Skeptics Side with Defendants," *New York Times*, August 8, 1990.
7. I am indebted to Rick Hornung for sharing with me his extensive files on the Central Park jogger case. See Rick Hornung, "The Central Park Rape: The Case against the Prosecution," *Village Voice*, February 20, 1990, 30–36.
8. "The O.J. Simpson Case-Discussion of Murder Case," *Ebony*, September 1, 1994, 37.
9. Thernstrom and Fetter, "From Scottsboro to Simpson," 17–27.
10. David Brooks, "The Duke Witch Hunt," *New York Times*, May 28, 2006; Duff Wilson, "Rape Accusation Has Ruined Lives, Students Say," *New York Times*, October 16, 2006; Peter Applebome, "Our Towns: After Duke Prosecution Began to

Collapse, Demonizing Continued," *New York Times*, April 15, 2007; Byron Calame, "Revisiting The Times's Coverage of the Duke Rape Case," *New York Times*, April 22, 2007.

11. John Steele Gordon, "Racial Role Reversal: What the Scottsboro Boys and the Duke Lacrosse Players Have in Common," *Wall Street Journal*, June 20, 2007.

12. Ibid.

Bibliography

Alabama State Planning Commission. *Alabama: A Guide to the Deep South*. New York: Richard R. Smith, 1941.

Allen, James S. "Organizing in the Deep South: A Communist's Memoir." *Nature, Society and Thought* 13, no. 3 (January 2001).

Anderson, Jervis. *Bayard Rustin: Troubles I've Seen. A Biography*. New York: Harper/Collins, 1997.

Andrews, William L. *To Tell a Free Story: The First Century of Afro-American Autobiography*. Urbana: University of Illinois Press, 1986.

Apel, Dora. *Imagery of Lynching: Black Men, White Women, and the Mob*. New Brunswick, NJ: Rutgers University Press, 2004.

Avery, Sheldon. *Up From Washington: William Pickens and the Negro Struggle for Equality, 1900–1954*. Newark: University of Delaware Press, 1989.

Baigell, Matthew, and Julia Williams, eds. *Artists against War and Facism: Papers of the First American Artists' Congress*. New Brunswick: Rutgers University Press, 1986.

Baker, Houston A., Jr. *Turning South Again: Re-Thinking Modernism/Re-Reading Booker T.* Durham, NC: Duke University Press, 2001.

Barson, Michael. *Better Dead Than Red!* New York: Hyperion, 1992.

Beals, Carleton. "Scottsboro Interview." Nation, February 12, 1936, 178–79.

———. "The Scottsboro Puppet Show." *Nation*, February 5, 1936, 149–50.

Bell, Bernard W. *The African American Novel and Its Tradition*. Amherst: University of Massachusetts Press, 1987.

Benstock, Shari. *Women of the Left Bank Paris, 1900–1940*. Austin: University of Texas Press, 1986.

Bernard, Emily, ed. *Remember Me to Harlem: The Letters of Langston Hughes and Carl Van Vechten, 1925–1964*. New York: Alfred A. Knopf, 2001.

Berry, Faith. *Before and Beyond Harlem: A Biography of Langston Hughes*. New York: Wings Books, 1995; orig. pub. Westport, CT: Lawrence Hill, 1983.

Beverly, William. *On the Lam: Narratives of Flight in J. Edgar Hoover's America*. Jackson: University Press of Mississippi, 2003.

Blakely, Allison. *Russia and the Negro: Blacks in Russian History and Thought*. Washington, DC: Howard University Press, 1986.

Bogdanovich, Peter. *Fritz Lang in America*. New York: Praeger, 1969.

Bogle, Donald. *Toms, Coons, Mulattoes, Mammies, and Bucks: An Interpretive History of Blacks in American Films*. New York: Continuum Publishing Company, 1989.

Bone, Robert. *The Negro Novel in America*. New Haven: Yale University Press, 1958; rev. 1965.

Bontemps, Arna. *Black Thunder*. Introduction by Arnold Rampersad. Boston: Beacon Press, 1992.

———. *The Old South*. New York: Dodd, Mead & Company, 1973.

Boyle, Kay. *Collected Poems of Kay Boyle.* Port Townsend, WA: Copper Canyon Press, 1991.

Brinkley, Douglas. *Rosa Parks.* New York: Viking, 2000.

Brown, Michael E., et al., eds. *New Studies in the Politics and Culture of U.S. Communism.* New York: Monthly Review, 1993.

Brown, Sterling A. *A Son's Return: Selected Essays of Sterling A. Brown,* ed. Mark A. Sanders. Boston: Northeastern University Press, 1996.

Brundage, W. Fitzhugh, ed. *Under Sentence of Death: Lynching in the South.* Chapel Hill: University of North Carolina Press, 1997.

Buchman, Frank N. D. *Remaking the World.* New York: Robert M. McBride, 1949.

Butler, Robert J., ed. *The Critical Response to Richard Wright.* Westport, CT: Greenwood Press, 1995.

Buttitta, Tony, and Barry Witham. *Uncle Sam Presents: A Memoir of the Federal Theatre, 1935–1939.* Philadelphia: University of Pennsylvania Press, 1982.

Cannon, Poppy. *A Gentle Knight: My Husband, Walter White.* New York: Rinehart, 1956.

Cappetti, Carla. *Writing Chicago: Modernism, Ethnography and the Novel.* New York: Columbia University Press, 1993.

Carby, Hazel V. "Ideologies of Black Folk: The Historical Novel of Slavery." In *Slavery and the Literary Imagination,* ed. Deborah E. McDowell and Arnold Rampersad (Baltimore: Johns Hopkins University Press, 1989), 125–43.

Carter, Dan T. *Scottsboro: A Tragedy of the American South,* rev. ed. Baton Rouge: Louisiana State University Press, 1979.

Carter, Herman J. D. *The Scottsboro Blues.* Nashville, TN: Mahlon Publishing Co, 1933.

Cash, W. J. *The Mind of the South.* New York: Alfred A. Knopf, 1941.

Ceplair, Larry, and Steven Englund. *The Inquisition in Hollywood: Politics in the Film Community, 1930–1960.* Garden City, NY: Anchor Press/Doubleday, 1980.

Chalmers, Allan K. *They Shall Be Free.* Garden City, NY: Doubleday and Company, 1951.

Chamberlin, William Henry. *Russia's Iron Age.* Boston: Little, Brown and Company, 1934.

Chambers, Whittaker. *Witness.* New York: Random House, 1952.

Chisholm, Anne. *Nancy Cunard: A Biography.* New York: Alfred A. Knopf, 1979.

Chura, Patrick. "Prolepsis and Anachronism: Emmet Till and the Historicity of *To Kill a Mockingbird.*" *Southern Literary History* 32, no. 2 (Spring 2000): 1–26.

Clurman, Harold. *The Fervent Years.* New York: Alfred A. Knopf, 1945.

Coiner, Constance. *Better Red: The Writing and Resistance of Tillie Olsen and Meridel Le Sueur.* New York: Oxford University Press, 1995.

Conroy, Jack, and Curt Johnson, eds. *Writers in Revolt: The Anvil Anthology, 1933–1940.* Westport, CT: Lawrence Hill and Company, 1973.

Cortner, Richard C. *A Mob Intent on Death: The NAACP and the Arkansas Riot Cases.* Middletown, CT: Wesleyan University Press, 1988.

Cotterill, Janet. *Language and Power in Court: A Linguistic Analysis of the O. J. Simpson Trial.* New York: Palgrave Macmillan, 2003.

Cowley, Malcolm. *The Dream of the Golden Mountain.* New York: Viking Press, 1980.

Craig, E. Quita. *Black Drama of the Federal Theatre Era: Beyond the Formal Horizons.* Amherst: University of Massachusetts Press, 1980.

Crenshaw, Files, Jr., and Kenneth A. Miller. *Scottsboro: The Firebrand of Communism.* Montgomery, AL: Brown Printing Company, 1936.

Crossman, Richard, ed. *The God That Failed.* New York: Harper, 1950.

Crowder, Henry (with Hugh Speck). *As Wonderful As All That? Henry Crowder's Memoir of His Affair with Nancy Cunard, 1928–1935.* Navarro, CA: Wild Trees Press, 1987.

Cruse, Harold. *The Crisis of the Negro Intellectual.* New York: William Morrow, 1967.

Cullen, Countee. *The Black Christ and Other Poems.* New York: Harper and Brothers, 1929.

——. *The Medea and Some Poems.* New York: Harper and Brothers, 1935.

Cunard, Nancy. *Negro: Anthology Made by Nancy Cunard, 1931–1933.* London: Wishart & Co., 1934; rpt. New York: Negro Universities Press, 1969.

Curridan, Mark, and Leroy Phillips, Jr. *Contempt of Court: The Turn-of-Century Lynching That Launched a Hundred Years of Federalism.* New York: Faber and Faber, 1999.

Curtin, Mary Ellen. *Black Prisoners and Their World, Alabama, 1865–1900.* Charlottesville: University Press of Virginia, 2000.

Davis, Allison. *Leadership, Love and Aggression.* New York: Harcourt, Brace, Jovanovich, 1985.

Davis, Angela Y. *Women, Race and Class.* New York: Vintage, 1983.

Davis, Benjamin J. *Communist Councilman from Harlem.* New York: International Publishers, 1969.

Davis, Frank Marshall. *Livin' the Blues.* Madison: University of Wisconsin Press, 1992.

Demby, William. *Bettlecreek.* Jackson: University Press of Mississippi, 1998.

Denning, Michael. *The Cultural Front: The Laboring of American Culture in the Twentieth Century.* New York: Verso, 1996.

D'Orso, Michael. *Like Judgment Day: The Ruin and Redemption of a Town Called Rosewood.* New York: G.P. Putnam's Sons, 1996.

Doyle, Laura. *Bordering on the Body: The Racial Matrix of Modern Fiction and Culture.* New York: Oxford University Press, 1994.

Dray, Phillip. *At the Hands of Persons Unknown: The Lynching of Black America.* New York: Random House, 2002.

Dudziak, Mary L. *Cold War Civil Rights.* Princeton, NJ: Princeton University Press, 2000.

Duffy, Susan. *The Political Left in the American Theatre of the 1930s: A Bibliographic Sourcebook.* Metuchen, NJ: Scarecrow Press, 1992.

Early, Gerald, ed. *My Soul's High Song: The Collected Writings of Countee Cullen, Voice of the Harlem Renaissance.* New York: Anchor Books, 1991.

Edgar, Robert R., ed. *An American in South Africa: The Travel Notes of Ralph J. Bunche, 28 September 1937–1 January 1938.* Athens: Ohio University Press, 1992.

Egerton, John. *Speak Now against the Day: The Generation before the Civil Rights Movement in the South.* New York: Alfred A. Knopf, 1994.

Elkins, Marilyn, ed. *Critical Essays on Kay Boyle.* New York: G. K. Hall & Co., 1997.

Endore, Guy. *Babouk.* New York: Monthly Review Press, 1991.

——. *The Crime at Scottsboro*. Hollywood, CA: Hollywood Scottsboro Committee, 1938.

Fabre, Michel. *Richard Wright: Books and Writers*. Jackson: University Press of Mississippi. 1990.

——. *The Unfinished Quest of Richard Wright*. New York: William Morrow and Company, 1973; 2nd ed., Urbana: University of Illinois Press, 1993.

——. *The World of Richard Wright*. Jackson: University Press of Mississippi, 1985.

Fairclough, Adam. *Better Day Coming: Blacks and Equality, 1890–2000*. New York: Viking, 2001.

Farnsworth, Robert M. *Melvin B. Tolson, 1898–1966: Plain Talk and Poetic Prophesy*. Columbia: University of Missouri Press, 1984.

Fenwick Library, George Mason University. *The Federal Theatre Project: A Catalog-Calendar of Productions*. Westport, CT: Greenwood Press, 1986.

Ferguson, Blanche. *Countee Cullen and the Negro Renaissance*. New York: Dodd, Mead & Company, 1966.

Foderaro, Lisa W. "Angered By Attack, Trump Urges Return of the Death Penalty." *New York Times*, May 1, 1989.

Foley, Barbara. *Radical Representations: Politics and Form in U.S. Proletarian Fiction, 1929–1941*. Durham, NC: Duke University Press, 1993.

Folsom, Franklin. *Days of Anger, Days of Hope: A Memoir of the League of American Writers, 1937–1942*. Niowot: University Press of Colorado, 1994.

Foner, Philip S., and Herbert Shapiro. *American Communism and Black Americans: A Documentary History, 1930–1934*. Philadelphia: Temple University Press, 1991.

Ford, Hugh, ed. *Nancy Cunard: Brave Poet, Indomitable Rebel 1896–1965*. Philadelphia: Chilton Book Company, 1968.

Fraden, Rena. *Blueprints for a Black Federal Theatre 1935–1939*. New York: Cambridge University Press, 1994.

Franklin, H. Bruce. *Prison Literature in America: The Victim as Criminal and Artist*, expanded ed. New York: Oxford University Press, 1989.

——. *The Victim as Criminal and Artist*. New York: Oxford University Press, 1978.

Frederickson, Mary. "Interview with Hollace Ransdall." November 6, 1974. Southern Oral History Project (#4007), Wilson Library, University of North Carolina at Chapel Hill.

Gayle, Addison. *Richard Wright: Ordeal of a Native Son*. New York: Doubleday, 1980.

Gill, Glenda E. *White Grease Paint on Black Performers: A Study of the Federal Theatre 1935–1939*. New York: Peter Lang, 1998.

Gilroy, Paul. *The Black Atlantic: Modernity and Double Consciousness*. Cambridge, MA: Harvard University Press, 1993.

——. *Small Acts: Thoughts on the Politics of Black Culture*. London: Serpent's Tail, 1993.

——. *There Ain't No Black in the Union Jack*. London: Century Hutchinson Ltd., 1987.

Goings, Kenneth W. *The NAACP Comes of Age: The Defeat of Judge John J. Parker*. Bloomington: Indiana University Press, 1990.

Golden, Thelma, ed. *Black Male: Representations of Masculinity in Contemporary American Art*. New York: Whitney Museum of American Art, 1994.

Goldstein, Malcolm. *The Political Stage: American Drama and the Theater of the Great Depression.* New York: Oxford University Press, 1974.

Goodman, James. *Stories of Scottsboro.* New York: Random House, 1995.

Gornick, Vivian. *The Romance of American Communism.* New York: Basic Books, 1977.

Grant, Joanne. *Ella Baker: Freedom Bound.* New York: John Wiley & Sons, 1998.

Graff, Ellen. *Stepping Left: Dance and Politics in New York City, 1928–1942.* Durham, NC: Duke University Press, 1997.

Green, Ben. *Before His Time: The Untold Story of Harry T. Moore, America's First Civil Rights Martyr.* New York: Free Press, 1999.

Green, Henry. *The Uncertainty of Everyday Life, 1915–1945.* New York: HarperCollins, 1992.

Greene, Lorenzo J. *Selling Black History for Carter G. Woodson: A Diary, 1930–1933,* ed. Arvarh E. Strickland. Columbia: University of Missouri Press, 1996.

Griffin, Farah Jasmine. *Who Set You Flowin'? The African-American Migration Narrative.* New York: Oxford University Press, 1995.

Griffiths, Frederick T. "Ralph Ellison, Richard Wright, and the Case of Angelo Herndon." *African American Review* 35, no. 4 (Winter 2001): 615–36.

Gunning, Sandra. *Race, Rape and Lynching: The Red Record of American Literature, 1890–1912.* New York: Oxford University Press, 1996

Hahn, Steven. *A Nation under Our Feet: Black Political Struggles in the Rural South From Slavery to the Great Migration.* Cambridge, MA: Harvard University Press, 2003.

Hale, Grace Elizabeth. *Making Whiteness: The Culture of Segregation in the South, 1890–1940.* New York: Vintage, 1999.

Hall, Jacquelyn Dowd. "Open Secrets: Memory, Imagination, and the Refashioning of Southern Identity." *American Quarterly* 50, no. 1 (1998): 109–24.

———. *Revolt against Chivalry: Jessie Daniel Ames and the Women's Crusade against Lynching.* New York: Columbia University Press, 1979.

———. "Women Writers, the 'Southern Front,' and the Dialectical Imagination." *The Journal of Southern History,* 69, no. 1 (February 2003): 3–38.

———. "'You Must Remember This': Autobiography as Social Critique." *Journal of American History* 85 (September 1998): 439–65.

Hammond, John, and Irving Townsend. *John Hammond on Record: An Autobiography.* New York: Ridge Press, 1977.

Hancock, Lynnell. "Wolf Pack: The Press and the Central Park Jogger Case." *Columbia Journalism Review* 41, no. 5 (January/February 2003): 38–42.

Hapke, Laura. *Labor's Text: The Worker in American Fiction.* New Brunswick, NJ: Rutgers University Press, 2001.

Harris, Joseph E. *African-American Reactions to the War in Ethiopia, 1936–1941.* Baton Rouge: Louisiana State University Press, 1994.

Harris, Trudier. *Exorcising Blackness: Historical and Literary Lynching and Burning Rituals.* Bloomington: Indiana University Press, 1984.

Haskins, James. *The Scottsboro Boys.* New York: Henry Holt and Company, 1994.

Hayden, Robert. *Heart-Shape in the Dust.* Detroit: Falcon Press, 1940.

Haywood, Harry. *Black Bolshevik: Autobiography of an Afro-American Communist.* Chicago: Liberator Press, 1978.

Healey, Dorothy, and Maurice Isserman. *Dorothy Healy Remembers: A Life in the American Communist Party.* New York: Oxford University Press, 1990.

Heilbut, Anthony. *Exiled in Paradise: German Refugee Artists and Intellectuals in America From the 1930s to the Present.* Berkeley: University of California Press, 1997.

Herndon, Angelo. *Let Me Live.* New York: Arno Press/New York Times, 1990.

Hicks, Granville, Joseph North, Michael Gold, et al., eds. *Proletarian Literature in the United States: An Anthology.* New York: International Publishers, 1935.

Higashida, Cheryl. "Aunt Sue's Children: Re-viewing the Gendered Politics of Richard Wright's Radicalism." *American Literature* 75, no. 2 (2003): 395–423.

Himelstein, Morgan Yale. *Drama Was a Weapon: The Left-Wing Theatre in New York, 1929–1941.* New Brunswick, NJ: Rutgers University Press, 1963.

Holcomb, Mark. "*To Kill a Mockingbird.*" *Film Quarterly* 55, no. 4 (Summer 2002): 34–40.

Hollis, Daniel Webster. *An Alabama Newspaper Tradition: Grover C. Hall and the Hall Family.* Tuscaloosa: University of Alabama Press, 1983.

Hughes, Langston. *The Big Sea.* New York: Alfred A. Knopf, 1940.

——. *The Collected Poems of Langston Hughes,* ed. Arnold Rampersad and David Roessel. New York: Alfred A. Knopf, 1994.

——. *Good Morning Revolution: Uncollected Writings of Social Protest by Langston Hughes,* ed. Faith Berry. Westport, CT: Lawrence Hill, 1973.

——. *I Wonder as I Wander: An Autobiographical Journey* (1956). New York: Thunder's Mouth Press, 1991.

——. *Scottsboro Limited: Four Poems and a Play in Verse.* New York: Golden Stair Press, 1932.

Humphries, Reynold. *Fritz Lang: Genre and Representation in His American Films.* Baltimore: Johns Hopkins University Press, 1989.

Hutchinson, Earl Ofari. *Beyond O.J.: Race, Sex, and Class Lessons for America.* Los Angeles, CA: Middle Passage Press, 1996.

——. *Blacks and Reds: Race and Class in Conflict, 1919–1990.* East Lansing: Michigan State University Press, 1995.

I'll Take My Stand: The South and the Agrarian Tradition. By 12 Southerners. New York: Harper and Row, 1962.

James, Winston. *Holding Aloft the Banner of Ethiopia: Caribbean Radicalism in Early Twentieth-Century America.* New York: Verso, 1998.

Janken, Kenneth Robert. *White: The Biography of Walter White, Mr. NAACP.* New York: New Press, 2003.

Jenkins, McKay. *The South in Black and White: Race, Sex, and Literature in the 1940s.* Chapel Hill: University of North Carolina Press, 1999.

Johnson, Charles S. *Shadow of the Plantation.* Chicago: University of Chicago Press, 1934.

Johnson, Claudia Durst. *Understanding "To Kill a Mockingbird."* Westport, CT: Greenwood Press, 2002.

Kaladjian, Walter. *American Culture between the Wars: Revisionary Modernism and Postmodern Critique.* New York: Columbia University Press, 1993.

Kanfer, Stefan. *A Journal of the Plague Years.* New York: Atheneum, 1973.

Kansas, Jane. "To Kill a Mockingbird." www.chebucto.ca/culture/mockingbird.

Kazin, Alfred. *On Native Grounds: An Interpretation of Modern American Prose Literature*. New York: Harcourt Brace, 1942.

——. *Writing Was Everything*. Cambridge, MA: Harvard University Press, 1995.

Kelley, Robin D.G. *Hammer and Hoe: Alabama Communists during the Great Depression*. Chapel Hill: University of North Carolina Press, 1990.

——. *Race Rebels: Culture, Politics, and the Black Working Class*. New York: Free Press, 1994.

Kempton, Murray. *Part of Our Time: Some Ruins and Monuments of the Thirties*. New York: Simon and Schuster, 1955; rpt. New York: Modern Library, 1998.

Kertesz, Louise. *The Poetic Vision of Muriel Rukeyser*. Baton Rouge: Louisiana State University Press, 1980.

Kester, Howard. *Revolt among the Sharecroppers*. Knoxville: University of Tennessee Press, 1997.

Kinnamon, Keneth. *The Emergence of Richard Wright*. Urbana: University of Illinois Press, 1972.

Kinnamon, Keneth, and Michel Fabre, eds. *Conversations with Richard Wright*. Jackson: University Press of Mississippi, 1993.

Kinshasa, Kwando Mbiassi. *The Man from Scottsboro: Clarence Norris in His Own Words*. Jefferson, NC: McFarland and Company, 1997.

Klehr, Harvey. *The Heyday of American Communism: The Depression*. New York: Basic Books, 1984.

Klehr, Harvey, John Earl Haynes, and Kyrill M. Anderson. *The Soviet World of American Communism*. New Haven: Yale University Press, 1998.

Klehr, Harvey, John Earl Haynes, and Fridrikh Igorevich Firsov. *The Secret World of American Communism*. New Haven: Yale University Press, 1995.

Klein, Marcus. *Foreigners: The Making of American Literature, 1900–1940*. Chicago: University of Chicago Press, 1981.

Langa, Helen. *Printmaking and the Left in 1930s New York*. Berkeley: University of California Press, 2004.

——. "Two Antilynching Art Exhibitions: Politicized Viewpoints, Racial Perspectives, Gendered Constraints." *American Art* 13, no. 1 (Spring 1999): 10–39.

Lee, Harper. *To Kill a Mockingbird*. New York: J.B. Lippincott, 1960.

Leibowitz, Robert. *The Defender: The Life and Career of Samuel S. Leibowitz, 1893–1933*. Englewood Cliffs, NJ: Prentice-Hall, Inc., 1981.

Lewis, David Levering. *When Harlem Was in Vogue*. New York: Alfred A. Knopf, 1984.

Little, Rivka Gewirtz. "Rage before Race: How Feminists Faltered on the Central Park Jogger Case." *Village Voice*, October 16–22, 2002.

Lubet, Steven, and Ann Althouse. "Reconstructing Atticus Finch." *Michigan Law Review* 97, no. 6 (May 1999): 1339–86.

Lumpkin, Grace. *Full Circle*. Boston: Western Islands, 1962.

——. *A Sign for Cain*. New York: Lee Furman, Inc., 1935.

——. *To Make My Bread*. Urbana: University of Illinois Press, 1995.

——. *The Wedding*. New York: Lee Furman, Inc., 1939.

Lumpkin, Katharine Du Pre. *The Making of a Southerner*. New York: Alfred A. Knopf, 1947; rpt. Westport, CT: Greenwood Press, 1971.

Mankiewicz, Don M. *Trial*. New York: Harper & Brothers, 1955.

Manning, Susan. "Black Voices, White Bodies: the Performance of Race and Gender in *How Long Brethren." American Quarterly* 50, no. 1 (March 1998): 24–46.

Marcus, Jane. "Bonding and Bondage: Nancy Cunard and the Making of the *Negro* Anthology." In *Borders, Boundaries, and Frames: Cultural Criticism and Cultural Studies*, ed. Mae Gwendolyn Henderson (New York: Routledge, 1995), 33–63.

Martin, Charles H. "Communists and Blacks: The ILD and the Angelo Herndon Case." *Journal of Negro History* 64, no. 2 (Spring 1979): 131–41.

Mathews, Jane DeHart. *The Federal Theatre, 1935–1939: Plays, Relief and Politics.* Princeton, NJ: Princeton University Press, 1967.

Maxwell, William J. *New Negro, Old Left: African-American Writing and Communism between the Wars.* New York: Columbia University Press, 1999.

McConachie Bruce A., and Daniel Friedman, eds. *Theatre for Working-Class Audiences in the United States 1830–1980.* Westport, CT: Greenwood Press, 1985.

McFeeley, William S. *Proximity to Death.* New York: W.W. Norton, 2000.

McGilligan, Patrick. *Fritz Lang: The Nature of the Beast.* New York: St. Martin's Press, 1997.

McGilligan, Patrick, and Paul Buhle. *Tender Comrades: A Backstory of the Hollywood Blacklist.* New York: St. Martin's Press, 1997.

McGovern, James R. *Anatomy of a Lynching: The Killing of Claude Neal.* Baton Rouge: Louisiana State University Press, 1982.

McWhorter, Ellen. "A Note on 'Communication to Nancy Cunard.'" www.english .uiuc.edu/maps/poetsa_f/boyle/cunard.htm.

Mellen, Joan. *Kay Boyle: Author of Herself.* New York: Farrar, Straus & Giroux, 1994.

Melli, Trisha. *I Am the Central Park Jogger.* New York: Scribner, 2003.

Miller, James A., Susan D. Pennybacker, and Eve Rosenhaft. "Mother Ada Wright and the International Campaign to Free the Scottsboro Boys, 1931–1934." *American Historical Review* 106, no. 2 (April 2001): 387–430.

Miller, Mark S., ed. *Working Lives: The "Southern Exposure" History of Labor in the South.* Chapel Hill: University of North Carolina Press, 1980.

Mitford, Nancy. *"The Pursuit of Love" and "Love in a Cold Climate": Two Novels.* New York: Modern Library, 1994.

Moon, Henry Lee. *Balance of Power: The Negro Vote.* Garden City, NY: Doubleday & Company, 1949.

Morrison, Toni, and Claudia Brodsky Lacour, eds. *Birth of a Nation'hood: Gaze, Script, and Spectacle in the O.J. Simpson Case.* New York: Pantheon Books, 1997.

Mostern, Kenneth. *Autobiography and Black Identity Politics: Racialization in Twentieth-Century America.* New York: Cambridge University Press, 1999.

Moynaugh, Maureen. *Nancy Cunard: Essays on Race and Empire.* Orchard Park, NY: Broadview Press, Ltd., 2002.

Mullen, Bill. *Popular Fronts: Chicago and African-American Cultural Politics, 1935–46.* Urbana: University of Illinois Press, 1999.

Mullen, Bill, and Sherry Linkon, eds. *Radical Revisions: Rereading 1930s Culture.* Urbana: University of Illinois Press, 1996.

Mullen, Harryette. *Freeing the Soul: Race, Subjectivity, and Differences in Slave Narrative.* New York: Cambridge University Press, 1999.

Murray, Hugh. "Changing America and the Changing Image of Scottsboro." *Phylon* 38, no. 1 (1st Qtr., 1977): 82–92.

———. "The NAACP versus the Communist Party: The Scottsboro Rape Cases, 1931–1932." *Phylon* 28, no. 3 (3rd Qtr., 1967): 276–87; rpt. in Bernard Sternsher, ed., *The Negro in Depression and War: Prelude to Revolution, 1930–1945* (Chicago: Quadrangle Books, 1969).

Naison, Mark. *Communists in Harlem during the Depression.* New York: Grove Press, 1983.

Navasky, Victor. *Naming Names.* New York: Viking Press, 1980.

Nelson, Cary. *Repression and Recovery: Modern American Poetry and the Politics of Cultural Memory, 1910–1945.* Madison: University of Wisconsin Press, 1989.

Nichols, Charles H., ed. *Arna Bontemps–Langston Hughes Letters, 1925–1967.* New York: Dodd, Mead and Company, 1980

Nolan, William A. *Communism versus the Negro.* Chicago: Institute of Social Order, 1950.

O'Connor, John, and Lorraine Brown, eds. *Free, Adult, Uncensored: The Living History of the Federal Theatre Project.* Washington, DC: New Republic Books. 1978.

Oshinsky, David. M. *"Worse Than Slavery": Parchman Farm and the Ordeal of Jim Crow Justice.* New York: Free Press, 1996.

Painter, Nell Irwin. *The Narrative of Hosea Hudson: His Life as a Negro Communist in the South.* Cambridge, MA: Harvard University Press, 1979.

Pasley, Fred. *Not Guilty! The Story of Samuel S. Leibowitz.* New York: G. P. Putnam's Sons, 1933.

Patterson, Haywood, and Earl Conrad. *Scottsboro Boy.* Garden City, NY: Doubleday & Co., Inc, 1950.

Patterson, William L. *The Man Who Cried Genocide: An Autobiography.* New York: International Publishers, 1971.

Peters, Paul, and George Sklar. *Stevedore.* New York: Covici-Friede, 1934.

Printz, Jessica Kimble. "Tracing the Fault Lines of the Radical Female Subject: Grace Lumpkin's *The Wedding.*" In *Radical Revisions: Rereading 1930s Culture,* ed. Bill Muller and Sherry Linkon (Urbana: University of Illinois Press, 1996), 167–86.

Rabinowitz, Paula. *Labor and Desire: Women's Revolutionary Fiction in Depression America.* Chapel Hill: University of North Carolina Press, 1991.

Rabkin, Gerald. *Drama and Commitment: Politics in the American Theatre of the Thirties.* Bloomington: Indiana University Press, 1964.

Rampersad, Arnold. *The Life of Langston Hughes.* Vol. I, 1902–1941: *I, Too, Sing America.* New York: Oxford University Press, 1986.

Ransdall, Hollace. *Report on the Scottsboro, Ala. Case,* made by Miss Hollace Ransdall, representing the American Civil Liberties Union, May 27, 1931. Unpublished ms.

Raper, Arthur F. *The Tragedy of Lynching.* Chapel Hill: University of North Carolina Press, 1933.

Record, Wilson. *The Negro and the Communist Party.* Chapel Hill: University of North Carolina Press, 1951.

Reilly, John. *Richard Wright: The Critical Reception.* New York: Burt Franklin, 1978.

Reynolds, Quentin. *Courtroom: The Story of Samuel S. Leibowitz.* New York: Farrar, Strauss and Company, 1950.

Rise, Eric W. *The Martinsville Seven: Race, Rape, and Capital Punishment.* Charlottesville: University Press of Virginia, 1995.

Robinson, Cedric J. *Black Marxism: The Making of the Black Radical Tradition.* London: Zed, 1983.

Robinson, Robert. *Black on Red: My Forty-Four Years inside the Soviet Union.* Washington, DC: Acropolis Press, 1988.

Rosengarten, Theodore. *All God's Dangers: The Life of Nate Shaw.* New York: Alfred A. Knopf, 1974.

Roux, Edward. *Time Longer than Rope: A History of the Black Man's Struggle for Freedom in South Africa.* Madison: University of Wisconsin Press, 1964.

Rowley, Hazel. *Richard Wright: The Life and Times.* New York: Henry Holt and Company, 2001.

Rukeyser, Muriel. *The Collected Poems.* New York: McGraw-Hill, 1978.

———. *The Life of Poetry.* New York: Current Books, 1949.

———. *Theory of Flight*, with foreword by Stephen Vincent Benet. New Haven: Yale University Press, 1935.

Salzman, Jack, and Barry Wallenstein, eds. *Years of Protest: A Collection of American Writings of the 1930s.* New York: Pegasus, 1967.

Saunders, Frances Stoner. *The Cultural Cold War: The CIA and the World of Arts and Letters.* New York: New Press, 1999.

Sayre, Nora. *Running Time: Films of the Cold War.* New York: Dial Press, 1982.

Schanberg, Sydney H. "A Journey through the Tangled Case of the Central Park Jogger." *Village Voice*, November 19, 2002.

Schechter, Patricia A. *Ida B. Wells-Barnett and American Reform, 1880–1930.* Chapel Hill: University of North Carolina Press, 2001.

Schrecker, Ellen. *Many Are the Crimes: McCarthyism in America.* Boston: Little, Brown and Company, 1998.

Shackleford, Dean. "The Female Voice in *To Kill a Mockingbird*: Narrative Strategies in Film and Novel." *Mississippi Quarterly* 50, no. 1 (Winter 1996/97): 101–14.

Shields, Charles J. *Mockingbird: A Portrait of Harper Lee.* New York: Henry Holt and Company, 2006.

Simon, Rita James, ed. *As We Saw the Thirties: Essays on Social and Political Movements of a Decade.* Urbana: University of Illinois Press, 1967.

Sitkoff, Harvard. *A New Deal for Blacks: The Emergence of Civil Rights as a National Issue.* New York: Oxford University Press, 1978.

Slingerland, Peter Van. *Something Terrible Has Happened.* New York: Harper & Row, 1966.

Smethurst, James Edward. *The New Red Negro: The Literary Left and African American Poetry, 1930–1946.* New York: Oxford University Press, 1999.

Smith, Lillian. *Killers of the Dream*, rev. and enlarged ed. New York: W.W. Norton & Company, 1961.

Smith, Valerie. *Self-Discovery and Authority in Afro-American Narrative.* Cambridge, MA: Harvard University Press, 1987.

Sollors, Werner. *Neither Black nor White Yet Both: Thematic Explorations of Interracial Literature.* New York: Oxford University Press, 1997.

Solomon, Mark. *The Cry Was Unity: Communists and African Americans, 1917–1936.* Jackson: University Press of Mississippi, 1998.

———. *Red and Black: Communism and Afro-Americans, 1929–1935.* New York: Garland Publishing, Inc., 1988.

Sowinska, Suzanne. Introduction to Grace Lumpkin, *To Make My Bread*. Urbana: University of Illinois Press, 1995.

Spanier, Sandra Whipple. *Kay Boyle, Artist and Activist*. Carbondale: Southern Illinois University Press, 1986.

Stepto, Robert B. *From Behind the Veil: A Study of Afro-American Narrative*. Urbana: University of Illinois Press, 1979.

Sternsher, Bernard, ed. *The Negro in Depression and War: Prelude to Revolution, 1930–1945*. Chicago: Quadrangle Books, 1969.

Stone, Irving. *Clarence Darrow for the Defense*. Garden City, NY: Doubleday, Doran and Company, 1941.

Stouffer, Samuel A. *Communism, Conformity, and Civil Liberties*. Garden City, NY: Doubleday & Company, 1955.

Stovall, Tyler. *Paris Noir: African Americans in the City of Light*. Boston: Houghton Mifflin Company, 1996.

Sullivan, Patricia. *Days of Hope: Race and Democracy in the New Deal Era*. Chapel Hill: University of North Carolina Press, 1996.

Sullivan, Timothy. *Unequal Verdicts: The Central Park Jogger Trials*. New York: American Lawyer Books/Simon & Schuster, 1992.

Sundquist, Eric J. "Blues for Atticus Finch: Scottsboro, Brown, and Harper Lee." In *The South as an American Problem*, ed. Larry J. Griffin and Don H. Doyle (Athens: University of Georgia Press, 1995).

Swados, Harvey, ed. *The American Writer and the Great Depression*. Indianapolis: Bobbs-Merrill Company, 1966.

Thernstrom, Abigail, and Henry D. Fetter. "From Scottsboro to Simpson." *Public Interest*, Winter 1996, 17–27.

Thompson, Frank T. *William A. Wellman*. Metuchen, NJ: Scarecrow Press, 1983.

Thurston, Michael. "Black Christ, Red Flag: Langston Hughes on Scottsboro." *College Literature* October 1995: 140.

Tindall, George B. *The Emergence of the New South, 1913–1945*. Baton Rouge: Louisiana State University Press, 1967.

Tolson, Melvin B. *Caviar and Cabbages: Selected Columns by Melvin B. Tolson from the "Washington Tribune," 1937–1944*, ed. Robert M. Farnsworth. Columbia: University of Missouri Press, 1982.

Troutt, David Dante. *The Monkey Suit and Other Short Fiction about African Americans and Justice*. New York: New Press, 1997.

Turner, Patricia A. *I Heard it through the Grapevine: Rumor in African American Culture*. Berkeley: University of California Press, 1993.

Turner, Patricia A., and Gary Alan Fine. *Whispers on the Color Line*. Berkeley: University of California Press, 2001.

Wald, Alan. "Between Insularity and Internationalism: The Lost World of the Jewish Communist Cultural Workers in America." In *Dark Times, Dire Decisions: Jews and Communism*, ed. Jonathan Frankel (New York: Oxford University Press, 2004).

———. *Exiles from a Future Time: The Forging of the Mid-Twentieth-Century American Left*. Chapel Hill: University of North Carolina Press, 2002.

———. *Trinity of Passion: The Literary Left and the Antifascist Crusade*. Chapel Hill: University of North Carolina Press, 2007.

———. *Writing from the Left: New Essays on Radical Culture and Politics.* New York: Verso, 1994.

Walker, Margaret. *Richard Wright: Daemonic Genius.* New York: Amistad, 1988.

Webb, Constance. *Richard Wright: A Biography.* New York: Putnam's, 1968.

West, Carroll Van. "Perpetuating the Myth of America: Scottsboro and Its Interpreters." *South Atlantic Quarterly* 80, no. 1 (Winter 1981): 36–48.

Wexley, John. *They Shall Not Die.* New York: Alfred A. Knopf, 1934.

White, Walter. *A Man Called White.* Athens: University of Georgia Press, 1995.

———. "The Negro and the Communists." *Harper's Monthly,* December 1931, 62–72.

Wiegman, Robyn. *Theorizing Anatomies: Theorizing Race and Gender.* Durham, NC: Duke University Press, 1995.

Wilkins, Roy, with Tom Mathews. *Standing Fast: The Autobiography of Roy Wilkins.* New York: Viking Press, 1982.

Williams, Linda. *Playing the Race Card: Melodramas of Black and White from Uncle Tom to O.J. Simpson.* Princeton, NJ: Princeton University Press, 2001.

Williamson, Joel. *The Crucible of Race: Black-White Relations in the American South Since Emancipation.* New York: Oxford University Press, 1984.

Wilson, Edmund. *The American Earthquake: A Documentary of the Twenties and Thirties.* Garden City: Doubleday & Company, 1958.

———. *The American Jitters: A Year of the Slump.* Freeport, NY: Books for Libraries Press, 1968.

———. *The Thirties: From Notebooks and Diaries of the Period.* New York: Farrar, Straus, Giroux, 1980.

Woodfin, Byron. *Lay Down with Dogs: The Story of Hugh Otis Bynum and the Scottsboro First Monday Bombing.* Tuscaloosa: University of Alabama Press, 1997.

Wright, Richard. "Blueprint for Negro Writing." *New Challenge* Fall 1937: 53–65; rpt. *Richard Wright Reader,* ed. Michel Fabre and Ellen Wright. New York: Harper, 1978.

———. *Early Works,* ed. Arnold Rampersad. New York: Library of America, 1991.

———. *Later Works,* ed. Arnold Rampersad. New York: Library of America, 1991.

Wright, Theon. *Rape in Paradise.* Honolulu: Mutual Publishing, 1990.

Young, James O. *Black Writers of the Thirties.* Baton Rouge: Louisiana State University Press, 1973.

Zangrando, Robert L. *The NAACP Crusade against Lynching, 1909–1950.* Philadelphia: Temple University Press, 1980.

Index